Les Roberts (Ed.)

Deep Mapping

MDPI

This book is a reprint of the Special Issue that appeared in the online, open access journal, *Humanities* (ISSN 2076-0787) from 2015–2016 (available at: http://www.mdpi.com/journal/humanities/special_issues/DeepMapping).

Guest Editor
Les Roberts
University of Liverpool
UK

Editorial Office
MDPI AG
Klybeckstrasse 64
Basel, Switzerland

Publisher
Shu-Kun Lin

Assistant Editor
Jie Gu

1. Edition 2016

MDPI • Basel • Beijing • Wuhan • Barcelona

ISBN 978-3-03842-165-8 (Hbk)
ISBN 978-3-03842-166-5 (PDF)

Table of Contents

List of Contributors

Laura Bissell: Contemporary Performance Practice, Royal Conservatoire of Scotland, 100 Renfrew St., Glasgow G2 3DB, UK.

James Cateridge: School of Arts, Oxford Brookes University, Gipsy Lane, Oxford OX3 0BP, UK.

Carenza Lewis: School of History and Heritage, University of Lincoln, Brayford Pool, Lincoln, LN6 7TS, UK.

Silvia Loeffler: Department of Geography, Maynooth University, Rhetoric House, Maynooth, County Kildare, Ireland.

Taien Ng-Chan: Centre for Interdisciplinary Studies in Society and Culture, Concordia University, 1455 de Maisonneuve Blvd. West, Montreal H3G 1M8, PQ, Canada.

David Overend: School of Media, Culture and Society, University of the West of Scotland, Ayr Campus, University Avenue, Ayr KA8 0SX, UK.

Claire Reddleman: Goldsmiths, University of London, Lewisham Way, New Cross, London SE14 6NW, UK.

Les Roberts: Department of Communication and Media, School of the Arts, University of Liverpool, 19 Abercromby Square, Liverpool L69 4ZG, UK.

Jos Smith: College of Humanities, The Queen's Drive, University of Exeter, Exeter, EX4 4QH, UK.

Giovanni Spissu: Department of Anthropology, Arthur Lewis Building School of Social Sciences, The University of Manchester, Manchester M13, UK.

Selina Springett: Media, Music, Cultural Studies and Communications, Macquarie University, North Ryde, Sydney 2109, Australia.

James Thurgill: School of Media, London College of Communication, University of the Arts London, London SE1 6SB, UK.

Denis Wood: Independent Scholar, 3205 Hillsborough Street, Raleigh, NC 27607, USA.

About the Guest Editor

Les Roberts lectures in the School of the Arts at the University of Liverpool. His research interests and practice fall within the areas of spatial anthropology, urban cultural studies, cultural memory, and spatial humanities. With a background in anthropology and cultural studies, his work explores the intersection between space, place, mobility, and memory with a particular focus on film and popular music cultures. He has published numerous peer-reviewed journal articles and book chapters on these and related subject areas, and is the author, editor or co-editor of a series of books on culture, space and place. These include Mapping Cultures: Place, Practice and Performance (2015), Locating the Moving Image: New Approaches to Film and Place (2014), Sites of Popular Music Heritage (2014), Liminal Landscapes: Travel, Experience and Spaces In-Between (2012), Film, Mobility and Urban Space: a Cinematic Geography of Liverpool (2012), and The City and the Moving Image: Urban Projections (2010).

Preface

Deep Mapping and Spatial Anthropology

Les Roberts

Abstract: This paper provides an introduction to the *Humanities* Special Issue on "Deep Mapping". It sets out the rationale for the collection and explores the broad-ranging nature of perspectives and practices that fall within the "undisciplined" interdisciplinary domain of spatial humanities. Sketching a cross-current of ideas that have begun to coalesce around the concept of "deep mapping", the paper argues that rather than attempting to outline a set of defining characteristics and "deep" cartographic features, a more instructive approach is to pay closer attention to the multivalent ways deep mapping is performatively put to work. Casting a critical and reflexive gaze over the developing discourse of deep mapping, it is argued that what deep mapping "is" cannot be reduced to the otherwise a-spatial and a-temporal fixity of the "deep map". In this respect, as an undisciplined survey of this increasing expansive field of study and practice, the paper explores the ways in which deep mapping can engage broader discussion around questions of spatial anthropology.

Reprinted from Humanities. Cite as: Roberts, L. Deep Mapping and Spatial Anthropology. *Humanities* **2016**, *5*(1), 5.

This Special Issue of *Humanities* on "Deep Mapping" came about as the result of an invitation to solicit and edit a collection of articles that address the state-of-the-art in scholarship on "spatial humanities"; but what *is* spatial humanities, exactly? Like its oft-related mothership descriptor "digital humanities", spatial humanities is a label which, while having at least more surveyable parameters than its parent territory, similarly finds itself in want of further exposition. Beyond offering what is at best a rather generic sketching of a disciplinary or interdisciplinary field of research in which concerns with "space" and "place" are understood to be in some way prominent, spatial humanities engages and incites curiosity precisely on account of what it does not succeed in detailing with any great precision. It always necessitates a "follow-through" of some description, even if this is merely to make the disciplinary or intellectual provenance that little bit more pronounced.

In much the same way that attempts to formalize a digital humanities discourse prompts the question as to what an "analogue humanities" might conceivably have looked like (a question I suspect no one in the pre-digital age much bothered themselves with), the discursive construction of a *spatial* humanities can be rendered critically (and playfully) askance by asking the question: what would, or does, a "temporal humanities" look like? Again, this a question I am not sure warrants much in the way of serious attention. Other than, perhaps, to imagine a scenario where, precipitated by an incipient "temporal turn" in humanities research, "time" suddenly becomes the rallying trope by which a re-constellation of otherwise diverse and disparate spheres of scholarship is brought into play. An unlikely turn of events, granted. And yet here we are with regard to space and the "spatial". Space has indeed become a rallying trope, and in this respect we could do a lot worse than look upon spatial humanities in a similarly aggregated fashion: *as a constellation of otherwise diverse and disparate spheres of scholarship*.

Keen to try and reflect the heterogeneity and plurality of research that falls within the (undisciplined) interdisciplinary domain of so-called spatial humanities, my interest was channelled towards exploring what is an undoubtedly more circumscribed subject area (or "sub-field", if you will), but one which is no less a product of this same rather loose configuration of spatio-cultural methods and practices. *Deep maps* and *deep mapping* offer particularly rich pickings in this respect in that they highlight the ways in which qualitative and humanistic forays into the representation and practice of space and place are multi-faceted, open-ended and—perhaps more contentiously—irreducible to formal and programmatic design. To flesh out the *doing* of deep mapping necessitates engaging with the same performative dynamics by which its various iterations are made manifest in practice. The concept is at best a convenient label to reach for when necessity demands but which quickly needs ditching the minute it conspires to mould itself into anything that starts to resemble a model or paradigm.

In this regard, it is instructive to look upon deep mapping in similar terms to those that define (and thwart) endeavours to formalize a discourse of "psychogeography". Which is not to suggest that the concepts are necessarily related (although there are obvious correlations and enticing points of overlap), but that they each represent labels that work best when they succeed in eluding containment or pat definition. Those comfortable with the tag "psychogeographer" would doubtless wince at attempts to shoehorn what it is they "do" into a kind of "how to..." guide: "Psychogeography for Beginners"; "An A-Z of Psychogeography" ("A is for Aragon", "D is for dérive", "S is for Situationist"), and so on. This instrumental approach to spatial humanities thinking is as problematic when it comes to deep mapping as it is for psychogeography. The elemental and very reasonably put question "what is deep mapping?" is best tackled not by outlining a set of defining characteristics and features (an exercise that is unavoidably weighted by the ballast of disciplinary persuasion) but by surveying the various precincts by which, as a coagulation of approaches and (inter)disciplinary interventions, it is performatively put to work. If, along the way, the sustainability or epistemological coherence of "deep mapping" (or of the "deep map", its artefactual product) is called into question then that itself may be a worthwhile and productive outcome of these proceedings.

But it equally well may be the case that the scrutiny and attention afforded to deep mapping in this special issue has the effect of providing a fresh set of insights by which otherwise different research practices may be tentatively brought into critical alignment. In this sense, deep mapping may be regarded as a statement of intent insofar as what it is *not* can at least be evinced and a certain familial resemblance correspondingly transacted. What it is not—or at least what it *should* not be—is irreducibly representational if by this we mean a process that is predicated on stemming the flow of spatial and temporal vitality that bleeds into and through the "map" (as a cartographic abstraction). It is on account of this necessarily processual underpinning to deep mapping practices that the very notion of a "deep map" becomes problematic. *PrairyErth*, its writer William Least Heat-Moon tells us, is a "deep map" [1]. But while I have no issue with the suggestion that the book may be the creative outcome of a process of "deep mapping", I am less sold on the idea that the text itself constitutes a "map". Although, as coiner of the term, Heat-Moon's name is routinely rehearsed in discussions of deep mapping as part of a preliminary conceptual backstory, Heat-Moon himself was not, of course, consciously laying the foundation stones for something that others would go on to pick up as "deep mapping". As a dense, "deeply" layered and richly textured literary survey of Chase County in the US state of Kansas, *PrairyErth* is a deep map of sorts; an entirely fitting metaphorical description of a textual cartography that aspires to yield what a conventional map or guide cannot even come close to conveying. What it is not is a representational device to which we can ascribe a set of formal and reproducible cartographic features that "project" Chase County or which provide a serviceable locative function (beyond that of a rudimentary stitching of narratives—however deep—to place). But, as ever with these things, it does kind of boil down to what is meant and understood by the term "map".

When we start to think about the ways *PrairyErth* may be considered a deep map there are a number of key touchpoints from which we can extrapolate a broader outline of analysis. As a self-styled "secretary of under-life" ([1], p. 367), Heat-Moon is desirous to dig deeper in his researches: to burrow down from the surface in order to *excavate* that which is hidden or buried beneath thinly-layered deposits of topsoil or asphalt. Deep mapping in this sense is as much a process of *archaeology* as it is cartography. With this comes an emphasis on *verticality* [2]: the "plumbing of a place's depth" ([3], p. 5). Horizontality is for the thin mappers ([4], pp. 29–31); those who hold back from peeling off the surface *layer* and who, in the process, thus allow limited space for *time*. The temporal configurations that anchor places in turf that has been synchronically as well as diachronically ploughed are the stuff from which the deep mapper fashions her or his craft. Our role (as readers, viewers, consumers, users) is to take up the invitation to "dive within", as artist, filmmaker, and transcendental meditator David Lynch might put it [5]. Wydeven writes that Heat-Moon "encourages us to fit ourselves in the creases [of maps]" ([6], p. 134), a nice turn of phrase which neatly captures the materiality and performativity that goes with the act of wayfinding: of exploring and placing oneself within the multi-scalar locative dimensions that are opened up through the *act* of deep mapping.

Another important touchpoint, one that casts a quizzical spotlight on the abstracted notion of a "deep map", is that deep map*ping* necessarily entails what Schiavini refers to as "deep travel" [2]. I do not wish to over labour the "deep" terminology here, but what Schiavini rather usefully points to is the performative work that goes into both the production of "deeply" configured spatial knowledge (what it is that the deep mapper "does") and what is precipitated by way of action performed *in response to* the production of such spatial knowledge. Were someone sufficiently inspired by *PrairyErth* (as the prototypical literary "deep map") to "dive within" the folds and creases of Chase County then they very well might find themselves tramping across the same geographical terrain that Heat-Moon's literary excavations have turned over. Deep mapping, in other words, cannot be reduced to the otherwise a-spatial and a-temporal domain of the (deep) *map*. It denotes an anthropology of practice. People are *doing things* when they engage in deep mapping; what it is they are doing becomes the focus of a spatial anthropology: a culture of mapping practice [7].

The important emphasis placed on *performance* is most notably explored by the archaeologists Mike Pearson and Michael Shanks, whose book *Theatre/Archaeology* [8] distils (by its title alone) a re-oriented and quintessentially interdisciplinary view of landscape, one that pays heed to "the grain and patina of place . . . the interpenetrations of the historical and the contemporary, the political and the poetic, the factual and the fictional, the discursive and the sensual" ([8], pp. 64–65). For Pearson and Shanks, deep mapping extends to "everything you might ever want to say about a place" ([8], p. 65). Of course, everything you might want to say may be voluminous, polyvocal or open-ended (as any deep mapping worth its salt should be aiming for anyway, almost by default). The loud thud that announced the arrival, by post, of *PrairyErth* on my doormat is testament to the ambition of its author to narratively and exhaustively scour the pocket of Kansas which the book sets out to "map". Unlike the surface dimensions that delineate and give shape to the locational properties of place, verticality and depth denote a comparative absence of limitations. The deeper you go the more layers you accrue. The problem becomes how to hold it all together: how to "frame" it as a map. The performativity and theatricality of place that might accompany a walker in Heat-Moon's Chase County, or which might give flavour to his or her practice, is not predicated on their being a material cartographic resource as a necessary reference point when out "in the field" (which is just as well given the book's size). The "map" is lodged in the more immaterial spaces of the body and imagination. Its performativity is made flesh in the way the walker inhabits and dwells within the space that both map (book) and walker conjure into being.

There is, then, a fundamental *creativity* at work in the practice of deep mapping, both on the part of the mapper and that of the "map reader". It is cartography as art rather than science (not that there is any neat binary between the two). Given this, it is not all that surprising to discover that, alongside the proponents of a literary deep mapping (chiefly, but by no means exclusively originating from the United States [9]), the most notable traffic of activity conducted under the banner of deep mapping has been initiated by visual and performance artists. Two of its most eloquent champions are Clifford McLucas and Iain Biggs [10–13]. The latter in particular is at pains to stress the interdisciplinarity or post-disciplinarity of deep mapping. For Biggs, one of the defining ingredients of an "open" deep mapping is the extent to which it is able to avoid "becoming complicit in its 'disciplining'" ([11], p. 21). This echoes the point made earlier about resisting the formalizing of a language or method of deep mapping that in some way reins it in as an otherwise "knowledgeable, passionate, polyvocal engagement with the world" ([11], p. 8). Cultivating what Biggs refers to as a *metaxy* of practice—a "space in-between" in which to pitch a precarious and purposefully indeterminate sense of a deep mapping practice—is to tread a fine line between complicity and creative dissolution. The creative efficacy of open deep mapping is co-extensive with that which underpins an artistic praxis that is operative outside of the tramlines of disciplinary or institutional orthodoxy. The complicity comes in the form of challenges that are posed in having to dance around a discursive object—deep mapping—whose constitutive "openness" is itself open to the dangers of "disciplining". In other words, the process of framing an "open" deep mapping runs the risk of a sort of "inverse disciplining" on account of the very fact that it *is* an object of discourse, even if it is trying its best not to be. The paradox is that Biggs's call for an "open deep mapping" only makes sense insofar as its openness is sufficiently diffuse as to do away with the very idea of deep mapping in the first place. The challenge of balancing these contradictory facets means questioning the coherence and validity of deep mapping on the one hand and maintaining a loose, plural and open application of the term on the other.

This careful balancing act is what I find myself having to be attentive to as editor of this special issue. "Deep maps" or "deep mapping" are not terms I have found myself using to any great extent in my work to date. Nor do I foresee a scenario where they are likely to imprint themselves more firmly on my thinking and practice. For me, deep mapping, like psychogeography, should be implicit not explicit in its application. As with my earlier observations with regard to "spatial humanities", deep mapping—resisting attempts to discipline what it is or should be, or mindful of its plurality and "openness" as a discursive frame of reference—might be better looked upon as a constellation of otherwise diverse and disparate spheres of scholarship. This is manifestly borne out in the disciplinary wide-ranging and eclectic nature of the contributions assembled in this special issue. Furthermore, as a gathering of multivalent expressions of deep mapping practice, there is a more formal and interactive rationale underpinning what digital and open-access publications such as this are able to offer the "deep map reader". That is, there is a more fluid and seamless interplay between the textuality of the writing and that of other media, whether these be photographic or moving images, digital maps, audio sound files, digital (and digitized) art works, locative media, hypertext data, other publications, and so on. Deep mapping, in short, is largely a product of the digital age.

This brings me back to where I came in: deep mapping situated in the wider context of an emerging digital/spatial humanities discourse. For deep mapping to acquire traction and resonance beyond an otherwise vague referencing to humanistic and qualitative approaches to the cartography of place—whether encompassing literary and cinematic geographies [14–17], psychogeography [18], site-specific art [19], popular music geographies and "musicscapes" [20,21], landscape and performance [8,22], spatial history [23], or whatever else we might wish to find room for in the big tent that is spatial humanities scholarship—then the possibilities offered by digital cultures and technologies certainly warrant attention.

The representational constraints attached to the idea of a deep map as something that aspires to be *more-than-representational* [24] are analogous to those that are routinely confronted by ethnographers tasked with the translation of *experience* (the flux and messiness of everyday life) into *narrative* (the ordered and disciplined fieldwork monograph). The fixity and abstraction of the cartographic frame (the map) belie the unboundedly complex, contingent and temporal spatialities of "the field". The deep map is a utopian imaginary of space inasmuch as it strives to frame or in some way open itself up to that which is "lived". By contrast, the *thin* map (if we can accept, for a moment, this oppositional conceit) is unapologetically representational: it is a representation of space that is ill- or, at least, under-equipped when it comes to servicing the needs of those whose inclinations are to "dive within". The writing culture debates that surfaced in anthropology in the 1980s, and which precipitated much hand-wringing in respect of a perceived crisis in ethnographic representation, drew closer attention to the interpretative mechanics of "thick description" in the writing-up of fieldwork "data". One of the consequences of this was to raise the question as to whether a sharply-observed and experientially immersive *literary* description of a given socio-cultural landscape could offer up as much if not more than a disciplinary-framed *ethnographic* account. A similar question could be posed in relation to cartography once the epistemological consequences of "depth" have been factored into the equation. If, as an exemplar of a geo-literary thick description, *PrairyErth* can be considered a "deep map" then might we not correspondingly draw the conclusion that a writer (or, indeed, filmmaker, artist, musician, or performer if we extend this to other branches of the arts) could be considered a deep cartographer on literary (or cinematic, artistic, musicological, or performative) terms alone? And, if so, doesn't this risk spreading what we might think of as the art of mapping just a little bit too thinly? Put another way: does "deep mapping" need to be discursively labelled as such for it to qualify as "deep mapping"? And if the answer is "no", then might not the cartographic hoops through which one might otherwise be required to jump be dispensed with altogether without any significant detraction in terms of what or how a place is being "mapped"?

These are questions I raise more by way of an introductory gesture than an attempt to render a partisan position as to the sustainability (or redundancy) of "deep mapping" as an object of discourse. There are certainly some common threads that can provisionally be woven together: a concern with narrative and spatial storytelling; a multi-scalar and multi-layered spatial structure; a capacity for thick description; a multimedial navigability; a spatially intertextual hermeneutics; an orientation towards the experiential and embodied; a strongly performative dimension; an embrace of the spatiotemporally contingent; a compliance with ethnographic and auto-ethnographic methods and frameworks; an "undisciplined" interdisciplinary modality; a time-based cartographics; an open and processual spatial sensibility; and, perhaps most telling, a reflexive—yet "aspirational" [25]—sense of the fundamental *unmappability* of the world the "deep map" sets out to map.

When we relate this all to developments in geo-spatial computing and the increasingly migratory domain of geographic information systems (GIS) then the idea of what a deep map might look (or act) like takes on more concrete form. Responding to the challenge to create a *model* of a deep map (already a questionable enterprise when appraised in light of my earlier misgivings) and to "explore how digital tools and interfaces can support ambiguous, subjective, uncertain, imprecise, rich, experiential content alongside the highly structured data at which GIS systems excel", Ridge et al conjure the notion of a "greedy deep map" ([26], pp. 176, 181). This rather intriguing and suggestive metaphor presents us with an image of a data-rich and data-hungry geospatial resource whose value lies in its capacity to outstrip the ability—and agency—of its human counterparts in terms of a spatial praxis sublimated towards more computational ends (the provision of a *potentiality* of retrievably layered data). To conceive of the deep map as "a space in which a near limitless range and quantity of sources can be included, interrogated, manipulated, archived, analysed, and read" ([26], p. 184) is to imagine what the realization of a deep map is or could be as a big data-driven, totalizing model. The question this raises for those invested in the development of a digital spatial humanities is whether the acquisition of the prized goal of a digitally limitless deep map comes at the cost of jettisoning the more anthropological, embodied and performative spatialities that are bound up with the practice of deep *mapping*.

Although, as geographer David Harvey observes, "maps are typically totalizing, usually two-dimensional, Cartesian, and very undialectical devices" ([27], p. 18), that does not, of course, mean that digital deep maps—or, rather, deep mapping practices that exploit the many possibilities and advantages offered by digital and geospatial technologies—are necessarily cut from the same Cartesian, undialectical cloth. As David Bodenhamer notes ([27], p. 23), at its best GIS-based deep mapping is an "ideal storyboard for humanists", offering a conceptual, technological, and spatial framework adapted to the need to tell spatial stories that are harvested from "experiential as well as objective space" and which are replete with the "rich contradictions and complexities" that ordinarily, as abstract representations of space [28], maps fall short of conveying.

Yet while the centrifugal pull of the digital world will continue to shape new ways of qualitatively mining the layered and experiential history of places, this should not be at the expense of a deep mapping praxis that is: (a) entirely at ease with the dispensing of programmatic labels (such as "deep mapping praxis"); (b) informed but not slavishly driven by digital tools and geospatial technologies; and (c) capacious enough to accommodate a diverse constituency of voices, actors, stakeholders, communities, and performers whose resonant clamour—the "multitonal chorus" ([9], p. 22) of everyday spatial dialogue—is not muted by the dead hand of corporate instrumentalism (as manifested by an increasingly audit- and impacts-driven culture of academic research). The tramping out of some form of common ground might be one way of approaching the breadth and diversity of the deep mapping on offer in this *Humanities* special issue. Another is to take heed of the broadly anthropological underpinnings that root deep mapping in the performative and processual flux of everyday life.

In their ethnographically-informed case study based in rural North Cornwall, Jane Bailey and Iain Biggs describe a deep mapping process that consists of "observing, listening, walking, conversing, writing and exchanging . . . of selecting, reflecting, naming, and generating . . . [and] of digitizing, interweaving, offering and inviting" ([29], p. 326). While this full set of verbs will not apply to all variations and permutations of deep mapping practice what they do usefully signpost is the way that very little of what deep mappers are *doing* is in fact oriented towards the production of maps so much as immersing themselves in the warp and weft of a lived and fundamentally intersubjective spatiality. It is from that performative platform—that space—that the creative coalescence of structures, forms, affects, energies, narratives, connections, memories, imaginaries, mythologies, voices, identities, temporalities, images, and textualities starts to provisionally take shape. Whether or not we wish to call what emerges from this process a "map" (or the process itself "mapping") seems to me less important than the fact that it is taking place at all. In its most quotidian sense, then, deep mapping can be looked upon as an embodied and reflexive immersion in a life that is lived and performed spatially. A cartography of depth. A *diving within*.

Conflicts of Interest

The author declares no conflict of interest.

References

1. William Least Heat-Moon. *PrairyErth (a Deep Map)*. Boston: Houghton Mifflin, 1991.
2. Cinzia Schiavini. "Writing the Land: Horizontality, Verticality and Deep Travel in William Least Heat-Moon's *PrairyErth*." *Revista di Studi Americani* 15–16 (2004–5): 93–113.
3. Christopher C. Gregory-Guider. "'Deep Maps': William Least Heat-Moon's Psychogeographic Cartographies." *eSharp*, no. 4 (2005), 1–17. Available online: http://www.gla.ac.uk/research/az/esharp/ issues/4/ (accessed on 23 December 2015).
4. Trevor Harris. "Deep Geography—Deep Mapping: Spatial Storytelling and a Sense of Place." In *Deep Maps and Spatial Narratives*. Edited by David J. Bodenhamer, John Corrigan and Trevor M. Harris. Bloomington: Indiana University Press, 2015, pp. 28–53.
5. David Lynch. *Catching the Big Fish: Meditation, Consciousness and Creativity*. London: Penguin, 2006.
6. Joseph J. Wydeven. "Review of PariryErth (a Deep Map)." *Great Plains Quarterly* 763 (1993): 133–34.
7. Les Roberts. "Mapping Cultures—A Spatial Anthropology." In *Mapping Cultures: Place, Practice, Performance*. Edited by Les Roberts. Basingstoke: Palgrave, 2012, pp. 1–25.
8. Mike Pearson, and Michael Shanks. *Theatre/Archaeology*. London: Routledge, 2001.
9. Susan Naramore Maher. *Deep Map Country: Literary Cartography of the Great Plains*. Lincoln: University of Nebraska Press, 2014.
10. Clifford McLucas. "Deep Mapping." 2000. Available online: http://cliffordmclucas.info/ deep-mapping. html (accessed on 23 December 2015).
11. Iain Biggs. "The Spaces of 'Deep Mapping': A Partial Account." *Journal of Arts and Communities* 2 (2010): 5–25.
12. Iain Biggs. "Deep Mapping as an Essaying of Place." 2010. Available online: http://www.iainbiggs.co.uk/textdeepmappingasanessayingofplace/ (accessed on 23 December 2015).

13. Iain Biggs. "'Deep mapping': A Brief Introduction." In *Mapping Spectral Traces*. Edited by Karen E. Till. Blacksburg: Virginia Tech College of Architecture and Urban Studies, 2010, pp. 5–8.
14. David Cooper, Christopher Donaldson, Patricia Murrieta-Flores, eds. *Literary Mapping in the Digital Age*. London: Ashgate, 2016.
15. Les Roberts. "Cinematic Cartography: Projecting Place through Film." In *Mapping Cultures: Place, Practice, Performance*. Edited by Les Roberts. Basingstoke: Palgrave, 2012, pp. 68–84.
16. Sébastian Caquard, D. R. Fraser Taylor, eds. "Cinematic Cartography." Special Issue, *The Cartographic Journal* 46 (2009): 16–23.
17. Julia Hallam, Les Roberts, eds. *Locating the Moving Image: New Approaches to Film and Place*. Bloomington: Indiana University Press, 2014.
18. Tina Richardson, ed. *Walking Inside Out: Contemporary British Psychogeography*. London: Rowman and Littlefield, 2015.
19. Alex Coles, ed. *Site-Specificity: The Ethnographic Turn*. London: Black Dog, 2001.
20. Sara Cohen, Robert Knifton, Marion Leonard, Les Roberts, eds. *Sites of Popular Music Heritage: Memories, Histories, Places*. Abingdon: London: Routledge, 2015.
21. Les Roberts. "Marketing Musicscapes, or, the Political Economy of Contagious Magic." *Tourist Studies* 14 (2014): 10–29.
22. David Crouch. *Flirting with Space: Journeys and Creativity*. Farnham: Ashgate, 2010.
23. Ian Gregory, Alistair Geddes, eds. *Towards Spatial Humanities: Historical GIS and Spatial History*. Bloomington: Indiana University Press, 2014.
24. Hayden Lorimer. "Cultural Geography: The Busyness of Being More-Than-Representational." *Progress in Human Geography* 29 (2005): 83–94.
25. Katie Oxx, Allan Brimicombe, and Johnathan Rush. "Envisioning Deep Maps: Exploring the Spatial Navigation Metaphor in Deep Mapping." *International Journal of Humanities and Arts Computing* 7 (2013): 201–27.
26. Mia Ridge, Don Lafreniere, and Scott Nesbit. "Creating Deep Maps and Spatial Narratives through Design." *International Journal of Humanities and Arts Computing* 7 (2013): 176–89.
27. David J. Bodenhamer. "Narrating Space and Place." In *Deep Maps and Spatial Narratives*. Edited by David J. Bodenhamer, John Corrigan and Trevor M. Harris. Bloomington: Indiana University Press, 2015, pp. 7–27.
28. Henri Lefebvre. *The Production of Space*. Oxford: Blackwell, 1991.
29. Jane Bailey, and Iain Biggs. "'Either Side of Delphy Bridge': A Deep Mapping Project Evoking and Engaging the Lives of Older Adults in Rural North Cornwall." *Journal of Rural Studies* 28 (2012): 318–28.

Going Deeper or Flatter: Connecting Deep Mapping, Flat Ontologies and the Democratizing of Knowledge

Selina Springett

Abstract: The concept of "deep mapping", as an approach to place, has been deployed as both a descriptor of a specific suite of creative works and as a set of aesthetic practices. While its definition has been amorphous and adaptive, a number of distinct, yet related, manifestations identify as, or have been identified by, the term. In recent times, it has garnered attention beyond literary discourse, particularly within the "spatial" turn of representation in the humanities and as a result of expanded platforms of data presentation. This paper takes a brief look at the practice of "deep mapping", considering it as a consciously performative act and tracing a number of its various manifestations. It explores how deep mapping is a reflection of epistemological trends in ontological practices of connectivity and the "flattening" of knowledge systems. In particular those put forward by post structural and cultural theorists, such as Bruno Latour, Gilles Deleuze, and Felix Guattari, as well as by theorists who associate with speculative realism. The concept of deep mapping as an aesthetic, methodological, and ideological tool, enables an approach to place that democratizes knowledge by crossing temporal, spatial, and disciplinary boundaries.

Reprinted from *Humanities*. Cite as: Springett, S. Going Deeper or Flatter: Connecting Deep Mapping, Flat Ontologies and the Democratizing of Knowledge. *Humanities* **2015**, *4*, 623–636.

1. Introduction

Both urban and rural spaces are saturated with stories. Every day we pass through these spaces we work, walk, live, and breathe them. Moreover, they are multi-textual and often highly politicized. Spectral traces of history ebb and flow in, through, and under the tide of contemporary life. To engage with these stories, this paper explores the use of "deep mapping" as a methodology and aesthetic choice. Deep mapping as an approach to place, aims to democratize knowledge through the crossing of temporal, spatial, and disciplinary boundaries. As a term and concept, has been used as both a descriptor of a certain type of approach to aesthetic representations of place (be they literary, performance based, or geo-representational), and a distinct set of aesthetic practices that can be linked, historically, to a number of diverse practitioners. More generally deep mapping can be categorized as involving intensive topographical exploration that aims to present diverse sources—histories, ecologies, poetics, memoires, and so on—as being of equally valid, and is often used to amplify the voices of marginalised stakeholders, both socially and ecologically. The aesthetic act of deep mapping as a practice, or set of practices is a method of creating a record of space, place or time that commits to an investment in enacting multi-vocal understandings: a "deep" (as opposed to shallow, one sided or perfunctory) investigation of place. "Deep map" first emerged as a literary term after being coined by American travel author William Least Heat Moon [1]. Moon spent nine years documenting Chase County, Kansas in the plains country of the Midwest United

States. In minute detail[1], he meticulously recorded and interwove interviews with locals, botanic information, Native American folklore and histories, literary and archival records, weather reports, geological data and cartographic references with travel writing and personal poetic reflections. As such, deep mapping has often been employed to engage with, narrate, and evoke multivocal, non-linear, open histories of place that are cross-referential. Opening up sometimes surprising resonances and dissonances.

The historical adaptations of deep mapping practices, from literary deep mapping, theatre archaeology, geographic information systems (GIS), and cross-disciplinary based productions, all strive towards more holistic methods of spatial representation. In perhaps its most common form, it is regarded as an intensive topographical research, encompassing spatial narratives, and with an aim to document, through the use of agency and inclusion, the interpenetrations of past and present [3]. I argue that the practice of deep mapping must be considered as a performative act, and one that can be perceived as a reflection of other concurrent ontological and epistemological trends discussed later in this paper. Karen Barad ([4], pp. 801–4) suggests a "*performative* understanding[2] of discursive practices challenges the representationalist belief in the power of words to represent pre-existing things." In a performative reading, according to Barad, the focus shifts from "questions of correspondence between descriptions and reality… to matters of practices/doings/actions." Deep maps go beyond description or simple communication, rather they are an *enaction* of place. They offer a certain type of storytelling that seeks to democratise knowledge,[3] through the use of *the map*. While this may not necessarily involve mapping in a traditional cartographic sense (although in some cases it does) deep mapping embodies the act of placing information on a plane of representation where the various components are connected metaphorically, and sometimes materially, by inhabiting the space on the same "map". As such, this mapping process attempts to give different knowledge equal audition or representation; be they botanical, historical, indigenous, folkloric or otherwise. Fundamentally, this seeks to be inclusive across fields and exemplifies an inherent interdisciplinarity.

The move towards a more explicit engagement between cartographic representations in geography through GIS and the arts has becoming steadily prevalent in discourses with arts and geography. Australian artists and academics Petra Gemeinboeck and Rob Saunders [5], in a critique of traditional cartographic exercises, speak of how the move to digital and GIS have given the illusion of a precise view of reality and suggest by engaging in art practice that apply alternative geographies it is possible to challenge this discourse. They propose that the "critical

---

1 The resulting mongraph is over 620 pages, rather long for a travel novel that covers an area 2015 km^2 with a recorded population in 2014 of just under 2700 [2].

2 Barad ([4], p. 801), in essence a new materialist, proposed a Posthuman understanding of performativity, that "allows matter its due as an active participant in the world's becoming, in its ongoing 'intra-activity'." While this argument is beyond the scope of this article it is interesting to consider when thinking through investigations of place.

3 By democratizing knowledge I mean that various knowledges are considered as of equal or important value in understanding of place; the folklore as much as the weather report; the local shopkeeper as much as the scientist; the river as much as the dam.

lenses of cultural, experimental and feminist geography distinguish themselves from cartographic science fiction by their desire for the embodied, multiple and plurivocal" ([5], pp. 160–62). In opting for this approach, they hope to challenge positivist notions of objectivity and truth. Scottish theorist and artist Iain Biggs ([6], pp. 5–9), who has written extensively on deep mapping, also draws on feminist theory, suggesting that deep mapping makes contributions to "a new ecology of embodied knowing" and should be seen in the form of "essaying" in the same way feminist reconstruction saw the essay as a "model of resistance". I read his work as asserting deep mapping as a method of production in which people can begin to see things in a relational way through underscoring the fundamental connectivity of various knowledge orders. The trend of eroding disciplinary boundaries leads geographer Daniel Sui ([7], pp. 62–64) to suggest that a "third culture" be created, one which embraces the traditional two culture model of arts and science, originally proposed by C.P. Snow [8] some fifty years earlier where analysis becomes a synthesis, "scientific rigor with artistic sensitivity, and pure intellectual pursuits with dominant societal concerns of our time". Although, that said, the authorizing knowledge production of science must be tempered in such statements.

This paper seeks to explore how deep mapping can be understood as echoing a trend in ontological practices of connectivity and subsequent epistemological "flattening" of knowledge. Specifically, systems put forward by poststructuralism and cultural theorists, such as Gilles Deleuze, Felix Guattari, and Bruno Latour, as well as those under the umbrella term of Speculative Realism and New Materialism including Levi Bryant, Timothy Morton, and Ian Bogost. This connection is both a reflection and a refraction, adopting in one sense the underlying theoretic drive but opting for a diverse spatiotemporal descriptor; one that is "deep" rather than "flat".

2. Trends of Production: Defining Deep Maps

The term deep mapping has been adapted or applied to a number of diverse projects and is becoming an increasingly popular as a signifier and an area of cultural production. Defined by Canadian literary academic Alison Calder ([9], pp. 164–70) early on as a type of "vertical travel writing", she explains, it interweaves "autobiography, archaeology, stories, memories, folklore, traces, reportage, weather, interviews, natural history, science, and intuition". The deep map has been adopted or reinterpreted to become both a methodological and philosophical approach driven by and extending into creative practice, including archaeological research, performance, GIS systems, and large scale art works. Pearson and Shanks([3], p. xi) and Calder ([9], p. 165) both suggest deep mapping blurs genres and while the former see it as involving the "recontexualisation of material" the latter emphasizes community as vital to the deep map. Calder ([9], p. 165) suggests the narrative of deep mapping as being 'cross-sections' which provide shifting and contingent readings of both human and natural landscapes. While her discussion focuses mostly on literary deep mapping, Mike Pearson [3] and Michael Shanks [10] write on a practice-based deep mapping, a type of environmental, ecological performance ethnography and a disciplinary practice described by Pearson and Shanks ([10], pp. 20–27) as "archaeological cultural poetics" that attempts to "record and represent the grain and patina of place". Both interpretations acknowledge that multiple and conflicting narratives connect and underscore this type of cultural production and that these narratives are equally important. It is of interest to note the close association between archaeology and deep mapping as being open to the politics of display and documentation and interrogating the perceived gap between subjectivity and objectivity. This recognition of such pre-existing hierarchies of knowledge and a desire to represent in a way that is more truthful of open multivocal contexts of place is an underlying current typical of deep mapping.

The theorization of both the performance and practice side of deep mapping coalesced through the development of a manifesto, which included ten tenets[4]. This arose from collaboration between the two directors of well-established and successful Welsh group *Brit Gof*, Clifford McLucas[5] [11,12] and Mike Pearson, and American archaeologist Michael Shanks [13]. The tenets themselves were jointly authored as part of a collaborative research project, called "*Three Landscapes*", sponsored and funded by Stanford University Humanities Center between 1999–2001. It was designed "to generate a creative short circuit between the artist's studio and the academy" [12]. These tenets were consequently adopted by Australian choreographer Rachael Swain [14] who, at the Australian Broadcasting Corporation Conference in 2012, quoted these ten points as being integral to the performances she undertakes as co-artistic director of the highly successful physical theatre company *Marrugeku*[6]. Her group utilises contemporary dance, circus skills, installation, video art as well as traditional and contemporary music in large-scale indoor and outdoor productions. Based

---

4 The tenets are as follows: (1) Deep maps will be BIG—the issue of resolution and detail is addressed by size. (2). Deep maps will be SLOW—they will naturally move at a speed of landform or weather. (3) Deep maps will be SUMPTUOUS—they will embrace a range of different media or registers in a sophisticated and multilayered orchestration. (4) Deep maps will only be achieved by the articulation of a variety of media—they will be genuinely multimedia, not as an aesthetic gesture or affectation, but as a practical necessity. (5) Deep maps will have at least three basic elements—a graphic work (large, horizontal or vertical), a time-based media component (film, video, performance), and a database or archival system that remains open and unfinished. (6) Deep maps will require the engagement of both the insider and outsider. (7) Deep maps will bring together the amateur and the professional, the artist and the scientist, the official and the unofficial, the national and the local. (8) Deep maps might only be possible and perhaps imaginable now—the digital processes at the heart of most modern media practices are allowing, for the first time, the easy combination of different orders of material—a new creative space. (9) Deep maps will not seek the authority and objectivity of conventional cartography. They will be politicized, passionate, and partisan. They will involve negotiation and contestation over who and what is represented and how. They will give rise to debate about the documentation and portrayal of people and places. (10) Deep maps will be unstable, fragile and temporary. They will be a conversation and not a statement.

5 Sadly, McLucas died in 2002 of a brain tumor, however, the manifesto, is accessible via a website set up posthumously by his friends and colleagues as a type of *momento mori*. His work continues to inspire and be used both as bases for productions and a starting point for many subsequent practice-led deep mapping projects. For a graphical example of a deep map produced by McLucas for large scale theatre archeology work *Tri Bywyd* (1995) see Kaye [11] and analysis of further works by *Brit Gof* complied under the umbrella of theatre/archeology in Shanks and Pearson [3].

6 *Marrugeku*'s work, which they explain as being a "process-driven, intercultural performance practice" [15] received far-reaching exposure in national and international arts festivals and has had a significant impact in raising awareness of Indigenous culture. Productions are created through long-term collaborations with artists from remote and urban locations, through international collaborations and in dialogue with Indigenous cultural custodians. They have been toured locally and internationally.

in Broome in the far north west of Australia, *Marrugeku*'s works explore intimate spatiotemporal stories through specifically indigenous and cross cultural collaborations and in consultation with community elders. Working closely with Kuwinjku artist and story keepers and the Yawuru people of Broome, the memories, tradition, stories, and lives of indigenous culture can be shared as can be seen in the highly successful *Mimi* (see Figure 1).

Figure 1. Production stills from Marrugeku production *Mimi,* Arnhem Land August/September 1998.

These proposed tenets of McLucas [10,11] adopted by Swain are useful in furthering my argument for both the performative nature of deep mapping—notably in the sense that it invokes a carrying out of something, as well as according to prescribed ritual—and in relation to how it functions as a democratisation of knowledge by exposing hierarchies.

The tenets of deep mapping as outlined on the *mometo mori* site of McLucas, have continues to be influential. Mapping Spectral Traces [16], of which McLucas was a member, is a transnational and interdisciplinary collective of artists and academics that work creating deep maps. This collective is also linked to a number of international creative, practice led, academic research centres called PLaCE [17], which "address issues of site, location, context and environment at the intersection of a multiplicity of disciplines and practices". There is no privileging or authorizing knowledge of one source of information over another and all agents have equal resonance in deep mapping, at least philosophically. The first point of the manifesto relates to the issue of resolution and states that deep maps should be "big". While this may not necessarily denote a physical size, the act of engaging in a deep map explicates a commitment to a large-scale investigation. The second tenet dictates that deep maps must be "slow" [12]. In this they call for an immersion in the subject that can only come with, and be actualised by time—not dissimilar to situated knowledge [18]. Deep maps [12], according to the tenets must" embrace a range of different media or registers in a…multilayered orchestration and may only be achieved by the articulation of a variety of media". This is certainly true of the work of *Marrugeku* [15]. According to tenet five deep maps will have at least three elements including a visual element, "a time-based media component…and a database or archival system that remains open and unfinished". With this he distinguishes in form from literary deep maps and while he lists as time based components film, video and performance—I would argue that sound, notably missing, should also be considered in this list. McLucas [12] then goes on to list as a necessity the inclusion of both privileged "insider" and the marginal "outsider", specifically of the "amateur and the professional, the artist and the scientist, the official and the unofficial, the national and the local". This strongly suggests an equal status of knowledge in the narratives of deep mapping.

McLucas [12] proposes deep maps are only now possible as different orders of materials may be easily combined within modern digital media practices. This is a discernable divergence from literary deep mapping. Whether this tenet is strictly true is debatable, as spatial representation can be manifested in numerous ways—not all necessarily digital. Although, that being said the popularity of GIS as a way of deep mapping must be noted. However, the penultimate tenet of McLucas' manifesto do reflect the sometimes political or ethical ideals underlying wider deep mapping practices, namely:

> Deep maps will not seek the authority and objectivity of conventional cartography. They will be politicized, passionate, and partisan. They will involve negotiation and contestation over who and what is represented and how. They will give rise to debate about the documentation and portrayal of people and places [12].

This is true of the work of PLaCE [17] who, as part of their mission, describe their work as focussing on the creation of a "supportive, open-ended space". They are interested in considering how they may engage, respectfully, in creative and research practices which employ "mapping" that seeks to "honour unacknowledged pasts and presences, and imagine more socially just futures." Their projects focus on employing visual and performing arts to address "such relevant concerns as ecological activism, place-based memory work, trauma, postcolonial geographies and related topics". Not unlike the underlying theme in the work of Least Heat Moon's [1] *Prarie Eryth* mentioned earlier which weaves historical, social, ecological, and indigenous narratives. Similarly, Rebecca Swain's collaborative work with co-director Dalisa Pigram [15] strongly ties in indigenous contestations in her performance theatre choreography and thematic explorations through works such as *Mimi* (see Figure 1), *Gudirr Gudirr* and *Cut the Sky* [15]. Now, let us consider one project of PLaCe that also works within an Australian context, *The Stony Rises Project* [19]. The presence of indigenous culture both in the past and in the present sense in a multi-tiered way is also poignantly included in this project. Run between 2008 and 2010, the work was expressed on manifold platforms including an artist camp, a travelling exhibition, a book, and in the community. Through the individual perspectives of a team of artists, scientists, designers, historians, curators and theorists,[7] they collectively created a deep map of a particular region (the Stony Rises near Lake Corangamite) within the Western District of Victoria, Australia. The investigation of one place and its features led to multiple histories being uncovered and shared.

This interpretation is in line with a more general rendering of what constitutes "deep mapping" and in particular can be seen as a way of de-colonizing. In each case traditionally prosaic modes of representation are combined with a particular place-conscious poetic that is socially and ecologically engaged. What I see as being clear from these examples is that creating a deep map is an act of *undoing*, a performative act that connects diverse disciplinary modes of enquiry and production, and blends ethics with aesthetics. The final tenet of McLucas alludes to the openness and humble nature of deep maps; rather than being a declaration or avowal they are to be considered a conversation. As such, various enactions of deep maps aim to present place as always open to the addition of supplementary voices, democratically positioning existent past, present and future knowledge and, thereby, building a structure of connectedness.

3. Mapping Ontologies of Connections and Flattened Epistemes

A recognition of connectedness in diverse epistemological approaches of deep mapping—which I perceive as abounding in a greater ontological trend of connectedness—traces pathways between the micro and the macro, the poetic and the prosaic, and past and present narratives. Let us begin by drawing comparisons with the work of French philosopher Bruno Latour [20,21]. I will deal first with his concepts outlined in his methodological paradigm: *i.e.* actor network theory (ANT). Through his ANT, Latour warns against neat "sizing". That is, nesting, or perhaps more clearly identifying micro processes within macro; a practice that according to Latour ([20], pp. 14–28) is a confusion of scale, and runs the risk of making leaps of cognition or oversimplifying causality. This could be equated with a top down systems logic, whereas Latour ([21], p. 175) states the micro should be examined *alongside* the macro, rather than as existing "above" or "below". Latour advocates slowing the investigative process down so as to trace the connections, which may have been previously considered "micro" to that which is "macro". The interactions and connections each actor makes are equally important in this system. This is echoed in the theoretical application of deep mapping, which calls for an unhurried investigation, a concerted effort to listen to and seek out multiple voices connected to place. Latour ([20], p. 15) warns against making assumptions or declarations without asking the all-important: "but why?" He suggests this is only achievable by slowing down and examining the minute interactions that connect the so-called micro with the macro. In this way it is possible to explore how existent human and nonhuman actors represent a kind of agency, and how these work in exchange with and on each other. In relation to deep

---

7 Run by the RMIT Design Research Institute it included, amongst others, artists Vicki Couzens, Lesley Duxbury, Ruth Johnstone, Seth Keen, Gini Lee, Jenny Lowe, Marion Manifold, Laurene Vaughan, Kit Wise, Carmel Wallace and Mark Minchinton; curators and writers Edmund Bernard Joyce, Dr. Heather Builth, Ross Gibson, Lisa Byrne, Harriet Edquist and Laurene Vaughan; and was coordinated by Laetitia Shand [19].

mapping it becomes an enaction of place in so much as that it puts into practice material explorations of the multitudinous exchanges that occur within such spaces. Perhaps most significant epistemological understanding here is the recognition of human and nonhuman actors as active rather than passive in ongoing mutual exchange processes. Concurrently, deep mapping processes map plurivocal agents that are to be considered as both connected and equal in import.

While Latour ([20], pp. 165–72) instructs to *flatten* the landscape or to *render it flat* in order to begin to understand the topography, he also, argues against "sizing" and "zooming" that give what he terms "false frames". By frames I understand him referring to conditions that impose restrictions to the possible connections or circumstances which may occur outside of the imposed frames. The flattening does not necessarily imply an elimination of depth then, but rather a democratisation of knowledge; for him, functionally, there is no pyramid of power. Latour ([20], p. 19) argues that "size and zoom should not be confused with connectedness". The use of what I refer to as *flat ontology* (a term initially borrowed from Manual De Landa [22], and further discussed by speculative realist Levi Bryant [23]), is something that is evident across a number of postmodern, poststucturalist and social constructivist discourses of the last half a century. It is a trope that argues against a hierarchical configuration. This may be understood as a democratisation of knowledge ordering. In this paradigm each "voice", be it historical, social or spanning diverse disciplinary understandings is equally audible or valid within the system. It points to a synchronous structuring with a focus on how the network is coordinated. This allows the importance of ongoing communications to be privileged, and the decentralisation of a single "objective" voice. This type of egalitarian connectivity can also be seen in the work of Deleuze and Guattari [24] who posit epistemes based on a rhizomatic structure or trope, eschewing the classic arboreal model—one which is inherently hierarchical—and enacts mapping rather than a tracing. In *Rhizome* Deleuze ([24], p. 12) writes "what distinguishes the map from the tracing is that it is entirely oriented toward an experimentation in contact with the real. The map does not reproduce an unconscious closed in upon itself; it constructs the unconscious". Rather than understanding being mapped neatly on a tree-like structure, a complex network of connections, each as important as the other, is made to explain ecologies, or complex systems of interrelatedness. Metaphorically, this is a flattened plain of connections. This also works to break down binaries of definition. No longer are we fixated on structured existing binaries of man [sic.] against nature, good *versus* evil, dominance *versus* submission, mind over body; rather what is to be highlighted is an intractable web of connectivity, inextricably bound together. Many theorists cited in this article—including Ian Bogost, Bruno Latour, and Graham Harman (amongst others)—argue that this dissolution of dualistic opposition is a reaction to the post-enlightenment or post Kantian philosophies, whose legacies are grounded in such dichotomies and hierarchical taxonomies. Similar to feminist theory, it argues against positivist notions of objective absolutism that are to be rejected. Tendencies are describes rather than causalities and personal narratives and testimonials are accepted as valid. Relatedly, feminist theorists Rick Dolphijn and Iris van der Tuin [25] understand New Materialism as being "transversal", specifically stepping away from dualism and representationalism to explore an active theory-making which interrogates paragdigms, spatio-temporality and the boundaries of disciplinarity. Interestingly, to tie into previous discussions connected to deep mapping Shanks ([26], p. 1) outlines archaeology as being distinctly materialist in terms seen as "a collection of what people do", rather than a "set of ideas or a body of knowledge", ideas which were formulated in dialogue with Latour, and Deleuze and Guattari whilst Shanks was in Paris.

Relevant to the modus of deep mapping—which is open to, or inclusive of, poetic, imagined or felt knowledge—are those of the Speculative Realists and Object Oriented Ontologists (OOO) [26]. Levi Bryant [23], Graham Harman ([27,28]), Ian Bogost [29], and Quentin Meillassoux [26], amongst others, bring into play the idea of "flat ontology". This is discussed in line with the theoretical assertion of this paper, the act of flattening knowledge systems can be read as a fundamentally democratizing action concurrent with the etho-ecological approach of deep mapping when applied to thinking about place. However, there is a distinct ethical or even political underpinning to most deep mapping projects that is missing from flat ontology. Particularly in the case of OOO, in which conceivably everything are deemed "objects" and nothing has special status—be it plate tectonics ([29], p. 9), an enchilada, the Taj Mahal, an iceberg ([29], p. 8), or digestion;—nonhumans, humans, and even ideas, for example, are all considered as being worthy of the same kind of philosophical metaphysical investigation. This assertion is to be understood not simply in a correlationist ([29], p.11) way—through their relation to humans—but as an equally important voice within discourse, rather than things being "elevated *up* to the status of humans" the whole system of objects are flattened to an egalitarian stasis. Not all loosely-bound members of OOO are strict adherents to a flat ontology [30]—Timothy Morton's ([30], pp. 1–24; [31]) posthuman "hyperobjects" for example, nuclear radiation, global warming, the Internet or the Earth are considered far greater in scale, time and, ultimately, importance than humans—there is still a flattening in the respect that the anthropocentric hierarchical model is abandoned. In such a system the centre of determination is not necessarily human but rather "object" oriented. The difference from some of these speculative realists to the theoretical models put forth by Deleuze and Guattari [24,32], and Latour [20] is that objects are presented as possessing unique qualities in and of themselves—that is, having properties unknowable to humans, and existing independently of humans. The emphasis is shifted to being on the connection they may have with both other non-human and human objects. However, where they do align, work towards shifting the metaphysical centre away from the singular human/ego, or even simple centralised node or hub of understanding. In this way denying the presence of a powerful nucleus or privileged standpoint from which all else can be defined is simply displaced or given realignment. Similarly, deep maps seek tend to deliberately dissipate the concept of a solid centre by actively generating pluralistic narratives. While the impetus to speculate flat ontological understandings inherent in OOO methods, there is a danger however, that by granting equal status to all objects you run the risk of losing any nuance of an increasingly necessary ecological (and ethical) argument. So while the electric massage chair, on a flattened ontological plane may have the same metaphysical complexities as the iceberg, do we offer the same amount of energy to debating or investigation of the demise of each? While humans, theoretically speaking, may be regarded as of no more importance than space dust, we are in a world where pressing social and environmental concerns require our action and thought. Similarly, while the idea of a flattened ontological ordering may on appearance be seen to be a democratising action, in effect, we might also detect an anti-humanist position, even if utopian. The power to flatten can appear to largely come, from a position of privilege. (The human is undoubtedly a privileged species). This must be seen as distinct to the democratising ideologies of deep mapping, which seeks to flatten hierarchies that are specifically social or ecological in nature.

Undeniably, there has been an ongoing, sometimes explorative, discursive shift away from an anthropocentric imagining of the world to one that is increasingly aware of the importance of non-human actors, both biological and ecological. I would propose, that the act of deep mapping has the potential to be inclusive of ecological as well as social concerns and work towards creating conversations that change the way people perceive, think about and ultimately engage with place. While this may not always be true of all the projects that carry the title of "deep map" it is something, which works within the framework of its ideology.

4. Deep Mapping the River

In my own work, I am seeking to explore the significance of a site through its permutations, and by the process of deep mapping in a consciously performative way. My own interpretation of deep mapping has evolved through engagement with an ongoing sound art project about the Cooks River, an urban river system in Sydney, Australia. The project offers a suite of works that use multiple sources including archival research, interviews with locals, and a range of stakeholders such as botanists, ecologists, environmental scientists, as well as collaborations with spoken word poets and field recordings of the river—both ambient and subaqueous. In these investigations, I have attempted to adhere to the philosophical and ethical drives of deep mapping by seeking out these multiple voices and research into geological and ongoing indigenous histories. These multiple layers coexist, to combine in different ways, media, and permutations in an attempt to create a deep map of the river.

To label a place as urban is often to discount the underlying topography of the urban/natural boundary. Liminal zones where natural and urban environments combine, and the stories of these zones, have often been shaped and scarred by an anthropocentric idea of the urban that is separate from its underlying natural ecology. This label can ignore the land it was built on, or the native flora and fauna that sometimes continues to share space with concrete and bitumen; and the rivers, creeks, and waterways that sustained life in the area for millennia have not always dried up, but can be found in the deeper strata, or as new diversions. Viewed in this way, the liminal status of urban rivers becomes not a deficiency; rather the stories of their becoming can be new and productive sites for exploration. Concomitant with Henri Lefebvre's [33] work on rhythm analysis, it is not merely the place which is significant, but how the site has been conceived of over the years, taking into account not only the *here and now,* but also rhythms that work over expanded time, not unlike the ebb and flow of the river. It would be impossible to investigate the Cooks River without taking into account the attitudes that shaped it over time, as collective imagined history moulded and continues to mould its identity and behaviour. The river in itself, a material signifier, is not simply the visible water but extends to include all the catchment. The gaze of humans over time has determined, to a large extent, the flow of the river and its ecological well-being. It has been viewed as water source, sacred country, sewer, drain, channel and is slowly returning to river. Through the enactment of a process of "deep mapping" the river, diverse and conflicting identities can be incorporated to create a rendering of place that aims to be both open, democratically located, inclusive of human and non-human actors and reflecting historic, narrative, and poetic imaginings. The following image (Figure 2) is from one of these projects made as part of a residency at the Bankstown Arts Centre and in collaboration with members of the Bankstown[8] Poetry Slam, who wrote and performed works inspired by ecological and historic stories of the river and their personal experiences. These words were then written literally into a large-scale mask of the river (see Figure 3). The mask was then removed, leaving the words behind echoing its shape and tributaries. The recorded works were then mixed into an accompanying sound piece that was installed into the exhibition space and experienced alongside live performances from the poets.

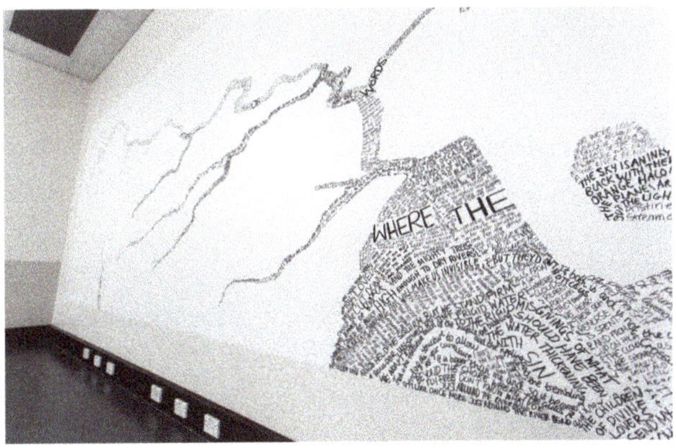

Figure 2. Installation still *Where the River Rises: A River of Words.* Created in collaboration with Bankstown Poetry Slam. Approx. 2.5m × 6m 2014 (Photo by Christopher Woe).

Figure 3. Installation still *Where the River Rises: A River of Words.* (Photo courtesy of artist).

---

8 Located in southwest Sydney Bankstown is one of the most culturally diverse local government areas in Australia with large diasporas of migrant populations. According to the last census in 2011 more than 85% of resident listed both parents born internationally and only 0.8% had both parents born in Australia [34]. The Bankstown government area is also the site where the Cooks River rises, a fact, many of the poets were unaware of prior to the project.

As touched on before this particular work forms only one layer of an ongoing palimpsest deep mapping project, it signifies the site where the river rises. It incorporates the tenets of deep mapping by working towards the inclusion of sometimes marginalized voices and poetic imaginations, and presents these as a recontextualisation of place. The performative enaction of this type of thinking about the river works to both locate and generate new and diverse imaginings in the collective stories generates *by/in the presence* of the river.

5. Conclusions

The drive to flatten or democratize knowledge systems, both in an epistemic and ethical sense, has slowly filtered into mainstream debates as seen in countless environmental and social justice campaigns. With the decline of an objectivist worldview, it is possible to distinguish a desire to give voice, or potentially listen more closely to those that may have been drowned out in the white noise of the contemporary mediascape. As both an artistic practice—or rather series of practices—and an ideological endeavour, deep mapping has the potential to incorporate an approach to spatio-temporal knowledge making that goes beyond more traditional forms of historic recreation, generating, in a sense, a conversation rather than the oration of one "objective historical voice". Shanks [35] applies a parallel reasoning presenting the work of archaeologists as an ecology of practices—a term he borrows form Isabelle Stengers [36]—aimed at mediating the material presence of the past. This approach and perspective offers alternative frameworks for combining creative practice and place conscious descriptions that through inclusive storytelling, create open interrogations of place. Deep mapping distinguishes itself from flat ontology by attempting to address such hierarchies, which are specifically social and ecological in nature in a way that is inherently political. It takes a performative approach of undoing or decolonizing, and seeks to affect a specific *enaction*, a process set into practice by altering or deepening peoples' perspective of place and their relationship to it—to one that is socially-just, ecologically-aware, more democratically-located, and one which erases disciplinary boundaries to extend beyond the academy.

Acknowledgments

I am indebted to Rebecca Swain for initially introducing me to the practice of deep mapping and for Virginia Madsen for her advice.

Conflicts of Interest

The author declares no conflict of interest.

References

1. William Least Heat Moon. *PrairyErth: (A Deep Map)*. New York: Houghton Mifflin Harcourt 1999.
2. The US Census Bureau. "Chase County QuickFacts." Available online: http://quickfacts.census.gov/qfd/states/20/20017.html (accessed on 20 September 2015).

3. Mike Pearson, and Michael Shanks. *Theater/archaeology*. London and New York: Routledge, 2001.
4. Karen Barad. "Posthumanist Performativity: Toward an Understanding of How Matter Comes to Matter." *Signs* 28 (2003): 801–31.
5. Petra Gemeinboeck, and Rob Saunders. "Urban fictions: A critical reflection on locative art and performative geographies." *Digital Creativity* 22 (2011): 160–73.
6. Iain Biggs. "The spaces of 'Deep Mapping': A partial account." *Journal of Arts and Communities* 2 (2011): 5–25.
7. Daniel Z. Sui. "GIS, Cartography, and the 'Third Culture': Geographic Imaginations in the Computer Age". *The Professional Geographer* 56 (2004): 62–72.
8. Charles Percy Snow. *The Two Cultures*. Cambridge: Cambridge University Press, 2012.
9. Alison Calder. "The Wilderness Plot, the Deep Map, and Sharon Butala's Changing Prairie." *Essays on Canadian Writing 77* (2002): 164–85.
10. Michael Shanks, and Christopher Y. Tilley. *Re-Constructing Archaeology: Theory and Practice*. New York: Routledge, 1992, pp. 97–100.
11. Nick Kaye. *Site Specific Art: Performance, Place, and Documentation.* New York: Routledge, 2000.
12. Clifford McLucas. "The Ten Tenets of Deep Mapping." Available online: http://cliffordmclucas.info/deep-mapping.html (accessed on 6 August 2015).
13. Michael Shanks. "The Three Landscapes Project." Available online: http://documents.stanford.edu/michaelshanks/62?view=print (accessed on 18 September 2015).
14. Rachael Swain. "Marrugeku and the ten tenets of deep mapping." Paper presented at Radio Beyond Radio ABC studios, Sydney, Australia, 7 September 2012.
15. Marrugeku Theatre. Available online: http://www.marrugeku.com.au/ (accessed on 10 August 2014).
16. Mapping Spectral Traces. Available online: http://www.mappingspectraltraces.org/ (accessed on 5 August 2015).
17. PLaCE International. Available online:http://placeinternational.org/index.htm (accessed on 14 October 2014).
18. Marcel Stoetzler, and Nira Yuval-Davis. "Standpoint theory, situated knowledge and the situated imagination." *Feminist Theory* 3 (2002): 315–33.
19. The Stony Rises Project. Available online: http://thestonyrisesproject.com/people/ (accessed on 7 March 2013).
20. Bruno Latour. "From realpolitik to dingpolitik." In *Making Things Public: Atmospheres of Democracy*. Edited by Bruno Latour and Peter Weibel. Karlsruhe and Cambridge: ZKM/Center for Art and Media in Karlsruhe and MIT Press, 2005, pp. 14–44.
21. Bruno Latour. *Reassembling the Social-An Introduction to Actor-Network-Theory*. New York: Oxford University Press, 2005, pp. 14–28.
22. Manuel DeLanda. *Intensive Science and Virtual Philosophy*. London and New York: A&C Black, 2013.
23. Levi R. Bryant. *The Democracy of Objects*. London: Open Humanities Press, 2011, pp. 1–25.

24. Gilles Deleuze, and Félix Guattari. *Rizoma (Introducción)*. Paris: Minuit, 1976, pp. 18–22.
25. Iris Van Der Tuin, and Rick Dolphijn. "The Transversality of New Materialism." *Women: A Cultural Review* 21 (2010): 153–71.
26. Michael Shanks. *Experiencing the Past: On the Character of Archaeology*. London: Routledge, 1992, pp. 15–45.
27. Ray Brassier, Iain Hamilton Grant, Graham Harman, and Quentin Meillassoux. "Speculative realism." *Collapse: Philosophical Research and Development* III (2007): 307–449.
28. Graham Harman. *Towards Speculative Realism: Essays and Lectures*. Hants: John Hunt Publishing, 2010, pp. 93–104.
29. Ian Bogost. *Alien Phenomenology, or, What it's Like to be a Thing*. Minneapolis: University of Minnesota Press, 2012, pp. 1–35.
30. Timothy Morton. *Hyperobjects: Philosophy and Ecology after the End of the World*. Minneapolis: University of Minnesota Press, 2013, pp. 1–20.
31. Timothy Morton. *Ecology without Nature: Rethinking Environmental Aesthetics*. Cambridge: Harvard University Press, 2009.
32. Felix Guattari. "The three ecologies." *New Formations* 8 (1989): 131–46.
33. Henri Lefebvre. *Rhythmanalysis: Space, Time and Everyday Life*. London and New York: Continuum, 2004.
34. Australian Bureau of Statistics. "2011 Census QuickStats." Available online: http://www.censusdata.abs.gov.au/census_services/getproduct/census/2011/quickstat/SSC10102?opendocument&navpos=220 (accessed on 23 September 2015).
35. Michael Shanks. *The Archaeological Imagination*. Walnut Creek: Left Coast Press, 2012.
36. Isabelle Stengers. "Introductory notes on an ecology of practices." *Cultural Studies Review* 11 (2013): 183–96.

Mapping Deeply

Denis Wood

Abstract: This is a description of an *avant la lettre* deep mapping project carried out by a geographer and a number of landscape architecture students in the early 1980s. Although humanists seem to take the "mapping" in deep mapping more metaphorically than cartographically, in this neighborhood mapping project, the mapmaking was taken literally, with the goal of producing an atlas of the neighborhood. In this, the neighborhood was construed as a transformer, turning the stuff of the world (gas, water, electricity) into the stuff of individual lives (sidewalk graffiti, wind chimes, barking dogs), and *vice versa*. Maps in the central transformer section of the atlas were to have charted this process in action, as in one showing the route of an individual newspaper into the neighborhood, then through the neighborhood to a home, and finally, as trash, out of the neighborhood in a garbage truck; though few of these had been completed when the project concluded in 1986. Resurrected in 1998 in an episode on Ira Glass' This American Life, the atlas was finally published, as Everything Sings: Maps for a Narrative Atlas, in 2010 (and an expanded edition in 2013).

Reprinted from *Humanities*. Cite as: Wood, D. Mapping Deeply. *Humanities* **2015**, *4*, 304–318.

Deep maps, deep mapping…

Yes, yes, but such a strange name for the practice. A practice that so often delivers far, far less than it promises. Especially maps. So many instances of deep mapping lack any at all.

Why mapping? Why not…thick description? Oh. Maybe because deep mapping is about *place*, while thick description is about…*behavior*? But aren't the two all mixed up together? Isn't that what deep mapping is supposed to be about—at least one of the things it's supposed to be about—the unfolding of human life *here*, the mutual relations of people and soil and plants and animals and…go on, you name it…*here* in this…place?

Well, obviously I'm just trying to figure out why they call it deep mapping, when mapping isn't what they are about, at all. They're storytellers mostly, which is great, but mostly they're not mappers. I'm talking about almost all of them, from William Least Heat-Moon to the most recent anthology of work on spatial narratives.

Not that you can't tell stories with maps. You can. In fact, every map tells a story, stor*ies* actually, many of them. Even maps that people who don't know much about maps call *thin maps* tell stories, ordinary, taken-for-granted maps, like highway maps, like the state highway map of North Carolina that John Fels and I spent fifty pages writing about back in 1986, and whose surface we barely scratched [1,2]. Thin maps…

Maps are models of concision, especially the ordinary taken-for-granted ones, cramming so many layers—so much history—into each line, into this line, for instance, this county border, the border of Wake County, first drawn in 1771 when the county was laid out of from parts of previously existing counties, but redrawn in 1787, 1881 and 1911, and named after Margaret Wake in 1771, the wife of William Tryon, then the colonial governor of North Carolina. All of this and *so much more* are caught up in that line that looks so simple but is anything but. And there're a *hundred* counties on this highway map of North Carolina, which also sports state borders, coasts, highways, roads, cities, towns, parks, reservations, military bases, forests and other things. This map is not simple, this map. It's not thin. It's deep and thick. Most maps are like this.

A lot of them wield power too, great power. We think about maps as being representations of the world, but they're not. They're arguments about the world, and many of these arguments are serious. "High court to hear map challenge in August" reads the headline to an article on the second page of yesterday's *News and Observer* [3]. A couple of days earlier, the lead editorial had been headed: "Rule on maps: the N.C. Supreme Court must quickly resolve a challenge to redistricting maps." [4]. These maps are about who gets to vote in which districts, that is, are about whether Democrats or Republicans will reign in state government. This has huge consequences for the distribution of wealth, education, health, you name it.

Let's not even think about the problems with immigration caused by the lines called national borders; or about the lines that bound school districts.

Some have more power than others, but all maps have it.

1. My Fight with Maps

My fight with maps, actually with cartography, was ignited by their rejection of modernism. As modernism was noisily turning its back on the failed rationalities, on the empty harmonies, on the make-believe coherences of Enlightenment, of Victorian thinking, cartography was clutching them ever more tightly to its breast. Painters may have been deconstructing pictorial space, composers shredding inherited tonalities, architects stripping walls of pilasters, cornices, and dentil moldings, poets following Pound's cry to "Make it new", and novelists indulging a self-consciousness that was all but the hallmark of the age, but cartographers, they were content to hone, to polish, to extend inherited forms.

Cartography exalted its unreflective empiricism as its *raison d'être*. It *cherished* the graphic conventions it had laid down in the 19th century. Even today, few maps acknowledge the 19th century's over. This, despite the fact maps were never what they were claimed to be, never what the map *themselves* claimed to be: veridical and value-free pictures of reality. They were always arguments about the way the maps' makers—or about the way those who paid the maps' makers—*thought* the world should be.

With modernism came a predisposition for resistance and smashing traditional forms, for going someplace stripped down, someplace essential, someplace real, for asking, *Why not?* I long felt around for a new map that wasn't of the same old subjects, that didn't have the same old forms, that looked and felt *modern.* Schoenberg wanted to emancipate the dissonance. Arp wanted to destroy existing modes of making art. Fifty years later, I wanted to destroy the existing ways of making maps through which millions were subjugated, herded, and all too often killed. I wanted to emancipate dream and desire as subjects of the map.

Hard to do in geography: it was nearly as hidebound as cartography.

But when I found myself teaching landscape architecture studios in the School of Design at North Carolina State University, I found my opportunity. I knew *nothing* about landscape architecture.

I knew less about studios, about how they worked, about what they were supposed to do. However, I figured landscape architects needed to know something about the environment in which they were working, and I figured that making maps might be a good way to learn—to discover—what it was they needed to know. So I set the first studio I taught—well, I set the *students*—the task of mapping a nearby neighborhood. The thing was, these were design students. They were *undergraduate* design students.

They had had little professional training (they weren't hidebound). They were wildly creative (which is why they had entered the School of Design). They knew *nothing* about the conventions of making maps (they were blank slates). So when I set them tasks like mapping sounds, or making maps from the perspective of bees, or constructing maps out of food they leaped at them like, like frolicking gazelles! They were all over these projects. They made the most amazing things.

I kept none of the maps. I mean, there were always more studios, more students, more maps. However, in a studio I co-taught with Robin Moore in the spring of 1982, we decided to make an atlas, a neighborhood atlas, an atlas we could reproduce on a copy machine, that we could distribute to the neighbors when we had finished. This meant the work had to make sense in black and white (in the early 1980s, color copy machines barely existed, and landscape architecture students *loved* to use colored markers), and it had to make sense to the neighbors (and so not be completely off-the-wall). This did not mean it had to be mapmaking the way these grad students had come to know it (and they were much more hidebound than the undergrads). *That* I was adamant about. But it didn't matter what they were mapping: I couldn't get them to leave the streets off their maps.

I was trying to get them to map the way the land smelled, the way it felt in their legs when they walked it, the way twilight made all the difference. I wasn't sure what the streets had to do with any of these, but the streets were an irreducible subject in the eyes of these students, the whatever-it-was that made the neighborhood a neighborhood. If you're going to be laying out subdivisions, which a lot of these students would be doing professionally, streets are really all you have to play with. I got that, but at the same time, the streets did seem to inhibit the other qualities I was trying to draw the students' attentions to. No matter how far into the background they intended the streets to recede, somehow they always stood out front.

Then, once when we were working on a map of streetlights, we just kept paring away the non-streetlights. We dumped the map crap (the neat line, the scale, the north arrow), the neighborhood boundaries, and the topography. Finally, we dumped the streets: first the scaled streets, then a schematic grid of the streets, finally even a hint of a grid of the streets. Daylight went too—that default daylight that most maps take for granted—so that we were fooling around with circles of white on a black background. That's when it became clear that the map wasn't about lamp *posts*, but about lamp *light*, and light was something we weren't sure how to deal with. Certainly, the uniform white circles we'd been drawing caught nothing of the way the light was fringed by the trees; and one night, armed with a camera, we scaled a fence and climbed a radio tower on the edge of the neighborhood hoping to catch the night lights on film. What a disappointment. The view from above was *nothing* like walking in and out of the pools of dappled light on the streets below. But I had a pochoir brush at home and when Carter Crawford—who had put himself in charge of atlas graphics—used it to draw the circles, it was magical (Figure 1). Nothing but blotches of white: that was the way it felt to be walking the streets at night.

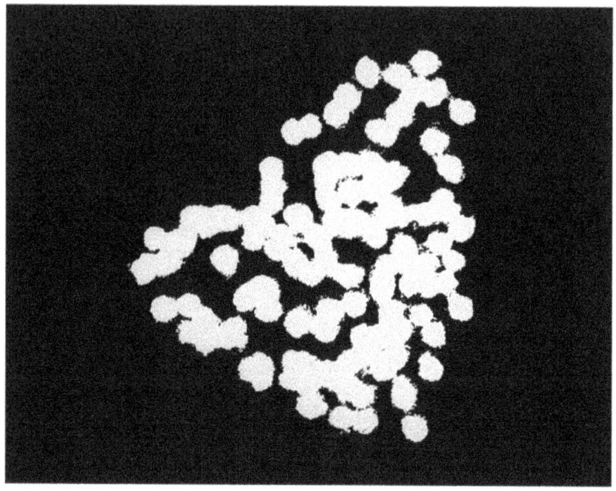

Figure 1. Pools of Light, map by Carter Crawford (from ([5], p. 53, used with permission).

The usual "efficient" map would have located everything on the street onto a single sheet—that is, different marks for lamp posts, fire hydrants, street signs, trees. Our *inefficient* map concentrated on a single subject and rather than lamp posts, it brought the pools of light into view. No legend, no north arrow, no neat line, none of the usual apparatus. At last: a modernist feel! Maybe even a sense of poetry, something imagistic, a little like Pound's "The apparition of these faces in the crowd;/Petals on a wet, black bough" [6] or Williams's red wheel barrow, but as it might manifest on a map, a map attentive to the experience of place [7].

That's when I knew we could write poems in maps. That's when I began thinking seriously about a poetics of cartography.

2. Making Maps

Once we got to this point, we started wanting to map everything. In the version of the atlas that was published in 2013 (Figure 2), there are 67 maps of the neighborhood [5], but back in the mid-1980s, we imagined well over a hundred, all the things we did make maps of—sewers and stars, streets and trees—but lots of others too, historical maps, the neighborhood as the Tuscarora would have known it, the neighborhood when it was a slave plantation, the neighborhood when it was laid out back in 1907, the neighborhood when my father had moved into it in 1921 and when he had moved out of it in 1927 (when he was six years old), the neighborhood after the soldiers came back from World War II, and…well, there were to have been a lot of historical maps. We wanted to map the history of the changes in lot ownership in regular increments (in the atlas as it stands there is a single map of ownership), the neighborhood gardens, selected block faces, the neighborhood as its kids knew it, the old-timers, the passers-through, the…

Figure 2. Cover of *Everything Sings* ([5], used with permission).

Well, almost no end to the list. The neighborhood mapping studios went on semester after semester, but students dropped out one after another—I mean, how long can anyone do this? Then, in 1986, I put it all in a box and forgot about it. It was nearly twenty-five years before it was published (though over the years it *did* acquire a certain notoriety).

All thats interesting, but not half as interesting as the forms the maps took as we took on the successive challenges the mapping threw at us. Figure 3 is the front endpaper from the second edition of *Everything Sings: Maps for a Narrative Atlas*. It was put together by Lisa Pearson, the publisher at Siglio (the press that brought the atlas out, first in 2010, then in a new edition with a bunch of new maps and other features in 2013), by excerpting segments from twelve of the book's maps. From top left, the first is from a map made by my father's older sister the time she visited us in the mid-1980s; the next from a map of the ages of the neighborhood's trees; the next from a map of house types; the next from a map of street signs; the next from a map of the postman's 1982 route; the next of the stars that shine on the neighborhood; the next of the neighborhood's fences; the next of a selection of radio waves passing through the neighborhood; the next of wind chimes; the next of property values; the next of the distance you can see out of the neighborhood from each of its intersections; and, the last, of the pumpkins that were on the porches, Halloween, 1982.

An equally varied bunch decorate the back endpaper (Figure 4): again, from upper left, footprints of the buildings in the neighborhood and surrounding area; the downtown Raleigh forest; the neighborhood's sewer, gas, and water lines; the topography of the hill the neighborhood tumbles over; the ages of the trees (a second time); a map of the flow of rent from the neighborhood; the power, telephone and cable lines; barking dogs; mentions in the neighborhood newsletter across its first decade; a map of sidewalk graffiti; of the sizes of the neighborhood trees; of its streetlights. No streets—though there *is* a map of streets in the atlas (along with one of traffic flows)—but the streets have such a profound affect on the location of everything else that you can pick them out on most of the maps, on the map of disfigured trees (Figure 5), for instance, which is more or less a map of where Carolina Power and Light had had the Asplundh Tree Expert Company butcher the trees to make sure the rare winter storms couldn't knock down the power lines along the streets and alleys.

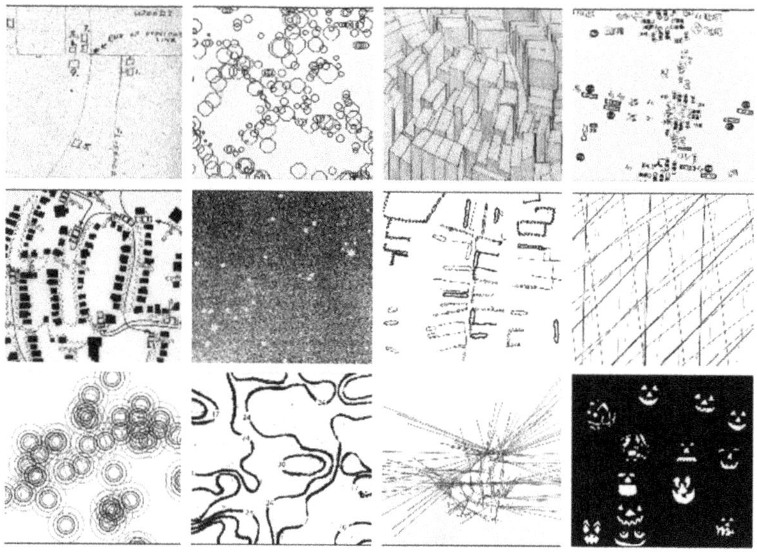

Figure 3. Front endpaper of *Everything Sings* ([5], used with permission).

The variation in these maps, which is characteristic of all the rest of them as well, reflects, of course, the work of individual students. Susan Waldrop's map of the neighborhood's fences (Figure 6) could have been made by no one else. Her map also reflects the wild variations in the data we collected, as well as our commitment to a poetics of cartography. Susan walked the neighborhood's streets and alleys to gather her data and, guided by our map of streetlights, laid it down like this (though she also had photos and rubbings). The straightforward way to make this map would have been to lay the fences down on a map of property lines (which would also be a map of streets). That way you could…what? What could you do with that map that you can't do with this, especially since elsewhere in the atlas there is a map of house numbers, and the two maps can be superimposed? In fact, this map of fences was one of the six maps published in the limited edition of *Everything Sings* as glicée prints on acid-free vellum, precisely so that you could superimpose them [8].

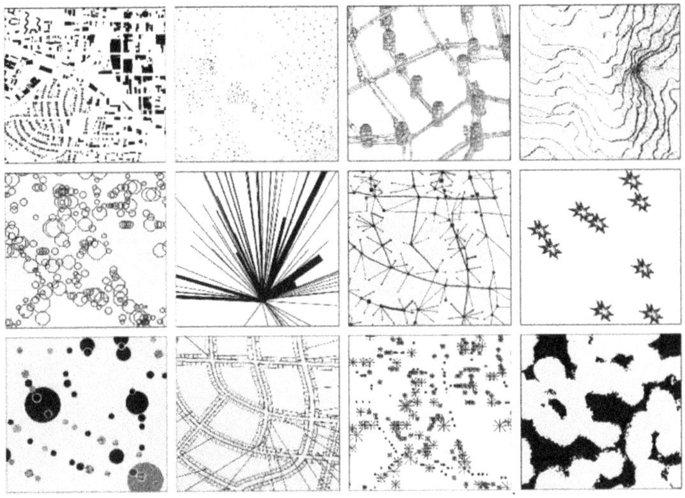

Figure 4. Back endpaper of *Everything Sings* ([5], used with permission).

Figure 5. Disfigured Trees, map by Shaub Dunkley ([5], p. 51, used with permission).

Figure 6. Fences, map by Helen Waldrop ([5], p. 95, used with permission).

As glicée prints, this limited edition included the maps of fences, overhead power lines, autumn leaves, police calls, graffiti, and wind chimes. Figure 7 shows us looking down through police calls (the numbers), overhead lines (the dots and lines), and fences. But if it had included the maps of the mains, the hill, the streets, the overhead power and other lines, large trees, and rooflines, you'd be able to look *down*, pretty deeply, from the tops of the houses through the trees, through the net of the power lines and the pavements to the storm drains below. Depending on the height of the roofline and the depth of the storm drain that could be fifty, sixty feet. That's pretty deep. Add the map of the stars and…

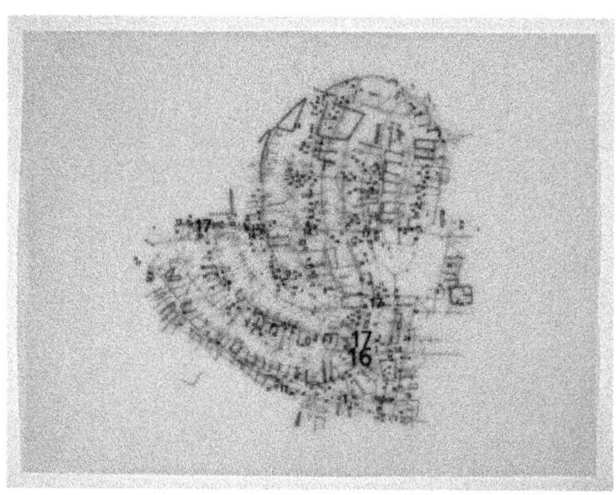

Figure 7. Three superimposed glicée prints, of Police Calls (data collected by numerous students, map by Denis Wood), Squirrel Highways (data collected by Shaub Dunkley, Carter Crawford and Denis Wood, map by Carter Crawford), and Fences (map by Helen Waldrop) ([8] used with permission).

3. Deep Mapping

But I doubt that's what deep mappers are talking about when they talk about deep mapping, though it would illustrate a meaning of the phrase worth thinking about. I've long wanted to make a map like it. On this map you'd look up at the neighborhood from below, from underneath the trees's deepest roots, up through that latticework—that mesh—to the mains, but then you'd look through the mains to the house connections snaking up *into* the houses and forking there to the toilets and sinks and tubs and showers like capillaries, and then down again, down the drains and through the waste pipes to the laterals and so down to the sewer lines, the house itself suspended in this web of flows, crystallizing out of them. *Can you see it?* You wouldn't see the house itself, just the water lines reaching up—as if to the sun, like branches—*almost* touching the drains. In the gap between? You, standing in the shower, the water shooting up from the underground, fountaining from the showerhead around you, cascading to the floor, pooling to the drain, and so down, down, down, you suspended in that gap, in that space, in that fountain. Lawrence Durrell says:

> You tell yourself that it is a woman you hold in your arms, but watching the sleeper you see all her growth in time, the unerring unfolding of cells which group and dispose themselves into the beloved face which remains always and for ever mysterious…And if, as biology tells us, every single cell in our body is replaced every seven years by another? At the most I hold in my arms something like a fountain of flesh, continuously playing, and in my mind a rainbow of dust [9].

Which is all the neighborhood is: a fountain of flesh, shingles, concrete, two-by-fours, trees, dirt, asphalt, iron pipes, starlight and the light cast through the leaves onto a summer night's sidewalk.

Which gets us a lot closer to what deep mappers have on their minds, the play of things and events that produce, that result in, that constitute the…neighborhood. At least in our case, the neighborhood. The name for the atlas originally—the one in the mid-1980s that we never published—was *Dancing and Singing*—from "Singing in the Rain", which is what we'd been doing one night, mapping in the rain—*A Narrative Atlas of Boylan Heights*. That was when we imagined we'd finish it. But of course we didn't, we couldn't. But when I published it as *Everything Sings: Maps for a Narrative Atlas*, though I changed everything else, I kept *narrative atlas*. "Maps *for*" acknowledges its unfinishability, but the retention of *narrative atlas* points to the fact that whatever else might have changed, its narrative structure—its narrative *intention*—remains the same.

Such a narrative could unfold any number of events in any number of ways but for us it had to unfold, first, our idea of what *neighborhoods did*, and then what *this particular neighborhood*, Boylan Heights, *actually did*. We made our idea of what neighborhoods did explicit, in introductions (the original had four of these), but all along our intention had been to lay it out through the example of our neighborhood, to let Boylan Heights itself speak for neighborhoods in general.

Our idea was this: the neighborhood is a process, a process-place or a process-thing that transforms anywhere into here, and here into everywhere, the city into the space of our lives, the citizen into the individual, and *vice versa*. Correspondingly, the atlas is organized in three phases that insensibly lead from one to the other. The first embodies the neighborhood's everywhere and anywhere quality, its *continuity* with the rest of the city; the second its character as a *transformer*, literally turning city stuff into neighborhood stuff (and *vice versa*); and the third, its irreducible uniqueness, its *discreteness* in the city. The first and third phases reflect each other through the transformer acting on them, so that if in the first phase the hill the neighborhood tumbles down is presented as a fact of geomorphology, in the third it shows up as the slopes the kids sled down in the snow (but we never mapped this sledding); or if in the first phase the neighborhood trees are just a part of the downtown Raleigh forest, in the third they acquire individuality, like the superlative water oak at 901 South Street.

Generic thing to unique thing: to make a pot of tea we light the flame on the stove. With the flame we pop—literally, actually, physically—up through space from the stove to the whole southeast of the United States, up through a nested hierarchy of spaces through a nested hierarchy of pipes of ever increasing size; from the slender tube running to the burner, to the pipe running to the stove, to the line running to the house, to the line running through the neighborhood, to the main snaking across the state, to the transcontinental pipeline tying together the many states of this southeast into one vast region of gas tied into the fields of the Gulf of Mexico. We are here; we are there; both at once; and the gas burns, time passes, and, over time, the heat latent in the gas is transferred to the water in the kettle and the water begins to boil. Then, the kettle whistles and, in time sings. In time and space, the tea is ready. I bring it to you on my porch, and we drink it, together, whole in Boylan Heights.

The maps and the text are at once very personal and yet essentially abstract, for while the atlas is very much about Boylan Heights, it's also about any neighborhood anywhere. They are *maps* with all of the science and technology that this implies, yet they have fingerprints all over them. I don't know where it comes from, but our maps have heart.

They also have trees and trees make another great example. The neighborhood trees are like neighborhood kids: they all grew up here. Of course, we also know that the trees are part of the greater Raleigh forest, which is but a piece of the great Southern Hardwood Forest which is…Suffice it to say, all the trees in the world are related. But this is to speak of trees *in general*. Once we relate the trees to their immediate environment (to a particular history of planting, nurturing, neglecting, cutting down—*i.e.*, to the saws of Asplundh), then we can begin to see patterns unique to Boylan Heights. We can see how unique an individual tree might be.

To get from one of these ends to the other? In the original idea of the atlas we'd intended for our map of Boylan Heights trees to be up front—sort of where it is the published atlas—and then to successively filter that map as we moved through the atlas. The last map would display a single, a unique tree. So first we recast the map of trees by size—we could have used age or something else—and then filtered out all but the large trees. This gave us a map of large trees. Then we recast the map of trees by species, and used oak as a filter to produce a map of oak trees (willow oaks, water oaks, Southern red oaks, Darlington oaks, and white oaks). We used this map to filter the map of large trees to give us a map of large oak trees. We then used our map of Public and Private Trees as a filter to get a map of large oaks in front yards. Filtering this by "good condition" and "location favorable to full growth" gave us nothing new, still all the large oaks in front yards. These we filtered for water oaks: large water oaks in front yards in good condition and a location favorable to full growth. This left us with four trees. We eliminated one of these that was suffering from Asplundh side-trimming to give us a map of large water oaks in front yards in *absolutely* good condition and a location favorable to full growth. Finally, filtering by age, we yanked the young and the old. This left us with large mature water oaks in front yards in absolutely good condition and a location favorable to full growth. Aha! There was only one of these, the magnificent water oak at 901 South Street. In the published atlas, where we printed all twelve filter maps on a single page, we turned its other side over to a photograph of this one great tree.

You can run this backwards. The neighborhood's a transformer: it doesn't care which way you go.

In this way it's like digestion, though not *quite* like digestion; though digestion *does* transform food, from without, into shit, from within, while producing life. In this way it works like a neighborhood. We tried to catch this aspect in a map that I think about as where the atlas pivots, from the early mapping of things coming in from outside the neighborhood—the maps of electricity, water, gas, and in this particular case, newspapers—to the maps of things the neighborhood produces—graffiti, the sounds of wind chimes, barking dogs, and in this particular case, garbage, trash. At the same time, in this instance, it generates conversation and shopping plans; consternation and amusement; movie choices and voting behavior (Figure 8).

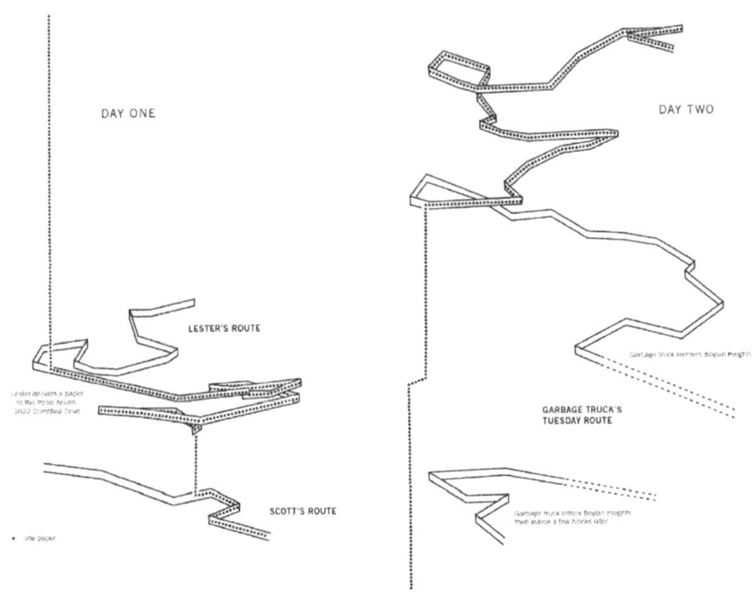

Figure 8. The Paper Route, data collected by Diane Pacella, Tim Hess, and Aurora Dee, map by Denis Wood ([5], p. 71, used with permission).

This space–time diagram—time running vertically up the page—comes after two other maps in the atlas. The first of these tracks Lester Mims on his paper route, the space–time ribbon of Lester delivering his route floating above an oblique map of the neighborhood, to which it's tethered at his house (where he picks the papers up that Scott has left him). The second map adds Scott's route.

Scott was the distribution manager for *Raleigh Times* District 170, the guy who left the bundle of newspapers on Lester's porch. His space–time ribbon is also floating above an oblique map of the neighborhood and, again, tethered at Lester's house. In Figure 8, we add to these a third space-time ribbon as we follow a Monday *Raleigh Times* through the neighborhood. This enters the neighborhood bundled in Scott's truck, is spread throughout Boylan Heights by Lester on his bicycle, and is then collected Tuesday morning by City of Raleigh garbage truck No. 1135, which trundles it off to the county landfill. The dotted line tracks the route of a single copy, digested (as something other than cellulose) by the Poole family members at 1022 Dorothea Drive. In the text facing the map we say, "The neighborhood: a metabolic machine (it eats newspapers)."

4. Coming to Deep Mapping

We didn't think about any of this as deep mapping. We had no idea what we were getting into, what we were doing. It started because I was asked to teach a landscape architecture studio and had no idea how to do it. The idea of having my students make maps was born of my background in geography and cartography, and the idea of having them map neighborhoods was born of my long-standing interest in neighborhoods. I'd brought the two together a couple of years earlier when I set high school students in Worcester, Massachusetts, the task of figuring out what teenagers meant when they said "neighborhood", and they had had other teenagers draw maps to help them figure it out [10]. I had written my master's thesis about teenagers' "mental maps" of their small town in southern Mexico where *barrios* were an essential piece of their identity [11], and my dissertation was about American teenage-tourists' "mental maps" of London, Rome, and Paris [12]. Maps and neighborhoods were things I thought about. However, having landscape architecture students get into them was sheer panic.

Since it turned out to be fun and productive, I'd done it several times. With different studios I'd done different neighborhoods and come at them with different questions but, as I said earlier, though I was getting more and more interested in the maps, I kept none of them. This was pretty much how we approached the atlas studio that spring in 1982: we were just going to make a bunch of maps and copy them into an atlas to hand out to the neighbors. The problem was…*atlas*? Atlases have structure, form. There's a progression among the maps in an atlas. They begin somewhere and they go somewhere else [13]. The students were picking things to map they were interested in or that they imagined they could find data for or they were just flailing around. But by mid-semester there *was* a list of maps that were going to be in the atlas and it just screamed…*structure*?

That was one thing. What really kicked us into gear, though, was the streetlight map, the streetlight map and the layout of the page it was going to be on. We had chosen 11″ × 17″ as our page size, opening to 11″ × 34″, and we were quite excited by the spread (Figure 9). We began to imagine that the atlas could be attractive—not just a jumble of maps—and this really pushed us to think about its structure. Clearly, this would have to have something to do with the neighborhood and this forced us to think—for the first time, two-thirds of the way through the semester—about the structure of the neighborhood. *Structure?* It forced us to think about neighborhoods, about what they were, how they functioned, what they did. We hadn't thought about this at all, just taken the neighborhood for granted. There was a lot of literature but it didn't take long to realize it wasn't going to help much—for many reasons—and one night—I don't remember how it happened—we had been talking it through and the idea of the transformer popped out, whole, like Venus arising fully formed from the sea.

This certainly gave us the structure we were looking for, but when we had laid out the maps, we had, or were going to have, it was like a few here and a few there. It was like a pamphlet. It was nothing like an atlas. The maps *were* hard to do. Much of the data had to be collected on the ground, walking through the neighborhood, systematically, again and again. Imagine Shaub Dunkley collecting the tree data: it took forever. It took way longer than the semester, and afterwards he had to code it for computer processing, *mainframe computer processing* (there were no real personal computers in 1982). Think about copying—nothing was online in those days—the maps for the water lines, the gas lines, the sewers. There were no small scale maps of these things: Aurora Dee and Tim Hess had to copy, and then compile the maps in the hands of the City of Raleigh and the Public Service Company of North Carolina; they had to reduce this mass to tractable form before they could even think about how they map would look. There weren't that many of us in the studio to begin with, and not everyone was as enthusiastic as me and Carter and Aurora and Shaub and Tim Hess and Jimmy Thiem.

Figure 9. Street Lights, by Carter Crawford ([14]; used with permission).

So we continued the work into the first summer semester, where the first thing we did was sketch out the maps we thought we'd need. This was when we began to think like deep mappers: what were the things that played a role? And then we had to figure out where to get the data or we had to go out and collect it ourselves. We began photographing, we began making rubbings, of everything. As we began to turn the new data into maps, and then into spreads, they pushed us to go out and collect more stuff. The mapping drove the thinking, drove the collecting, drove the design; and all these things drove the mapping, pushing us into new subjects, forcing us to find or collect new data, and...it didn't stop. The first summer semester bled into the second; and that bled into the fall semester and that into the spring. That spring studio was the fifth, and it was the last formal atlas studio for which credit was given.

It was nuts and gradually it withered away, but it's important to acknowledge how the mapping drove the project forward. It wasn't a project for which some maps were made for illustrative purposes. It was a project in which the mapping insisted on the collecting of data and the asking of questions and the drafting of maps, which in turn insisted on the collection of further data, on the asking of further questions, on the drafting of further maps. This never really ended. A few years later, I published *The Power of Maps* to accompany a show I'd curated for New York's Cooper-Hewitt, the Smithsonian's design museum (a couple of years later we remounted it at the Smithsonian in Washington) [15]. The shows and the book were popular, and they led to articles in *Scientific American* and *Cartographica* and other magazines and journals [16], including the recent collection, *Rethinking Maps* [17]. There *are* a couple of the atlas maps in *The Power of Maps* but what led a producer from Ira Glass' *This American Life* to set up an interview for him to talk with me were…maps in general, on background, for a show he was planning on maps. At the end of the hour he asked me if I made maps myself and I said, "No, not really, I'm more of a map theoretician, though I *have* been working on this atlas of my neighborhood for a number of years." He asked about it and decided we needed to talk for another hour.

Suddenly the interview wasn't on background anymore.

The broadcast of that episode of *This American Life* changed everything. Shortly thereafter I was approached by a curator for the Tang Teaching Museum at Skidmore College who wanted to know if I'd be interested in exhibiting any of the maps in a show of map art he was putting together [18]. He wanted the draft maps, the *art*. He was the first of many. Shortly after that, publishers began getting in touch. My favorite was the big art publisher who, in rejecting the project, told me to never again send black-and-white photocopies of the work. Apparently she hadn't gotten the message that it was a black and white book. Twelve years later, siglio brought *Everything Sings* out with an introduction by Ira Glass and a dozen new maps made from our old data; and then three years later brought out a second edition with another ten new maps. There's still tons of data. A Korean edition came out earlier this year [19]. It's a juggernaut.

5. Concluding Remarks: Mapping Deeply

I grant the idiosyncrasies of the project—people cannot get enough of the pumpkin map—and the attractiveness of the maps, especially their variety; but it's worth stressing that the key is…*mapping*; and by mapping I don't mean dropping some data into a computer mapping program, I mean getting out and doing the fieldwork. I also grant that Boylan Heights is small and therefore comparatively tractable, but what's essential *is getting out in the field*—whatever that field is—and looking hard at stuff. Walking through it and writing it down forces a valuable kind of attention, an irreplaceable kind of attention.

This kind of immersion makes you think about things, dream about them, and this prompts new questions, which send you back out into the field. It doesn't take long to get deep into things when you're paying attention, and mapping focuses attention. I know I've said this all before, but if I thought it would help, I'd say it again. And again. *Get deep into things*!

Conflicts of Interest

The authors declare no conflict of interest.

References

1. Denis Wood, and John Fels. "Designs on Signs: Myth and Meaning in Maps." *Cartographica* 23 (1986): 54–103.
2. Denis Wood, and John Fels. "Designs on Signs: Myth and Meaning in Maps." In *Classics in Cartography: Reflections on the Most Influential Articles from the Journal Cartographica.* Edited by Martin Dodge and Jeremy Crampton. New York: Wiley, 2011, pp. 209–70.
3. Anonymous. "High court to hear map challenge in August." *News and Observer*, 8 May 2015, p. A2.
4. Anonymous. "Rule on maps: The N.C. Supreme Court must quickly resolve a challenge to redistricting maps." *News and Observer*, 6 May 2015, p. A12.
5. Denis Wood. *Everything Sings: Maps for a Narrative Atlas*, 2nd ed. Revised and Expanded. Los Angeles: Siglio, 2013.
6. Ezra Pound. "In a Station of the Metro." *Poetry*, April 1913, p. 12.
7. William Carlos Williams. "XXII." In *Spring and All.* New York: Contact Publishing, 1923.
8. Denis Wood. *Everything Sings: Maps for a Narrative Atlas.* Los Angeles: Siglio Editions, 2010.
9. Lawrence Durrell. *Clea.* New York: Dutton, 1960, p. 98.
10. Denis Wood. "A neighborhood is to hang around." *Children's Environments Quarterly* 1 (1984–1985): 29–45.
11. Denis Wood. *Fleeting Glimpses, or Adolescent and Other Images of the Entity Called San Cristobal las Casas.* Worcester: Clark University Cartographic Laboratory, 1971.
12. Denis Wood. *I Don't Want To, But I Will.* Worcester: Clark University Cartographic Laboratory, 1973.
13. Denis Wood. "Pleasure in the Idea: The Atlas as Narrative Form." *Cartographica* 24 (1987): 24–45.
14. Denis Wood. *Dancing and Singing: A Narrative Atlas of Boylan Heights.* Unpublished manuscript, 1982.
15. Denis Wood. *The Power of Maps.* New York: Guilford Press, 1992.
16. Denis Wood. *"The Power* of Maps." *Scientific American* 268 (1993): 88–83.
17. John Krygier, and Denis Wood. "Ce n'est pasle monde (This is not the world)." In *Rethinking Maps: New Frontiers in Cartographic Theory.* Edited by Martin Dodge, Rob Kitchin and Chris Perkins. New York: Routledge, 2009, pp. 188–19.
18. Susan Bender, and Ian Berry. *The World According to the Newest and Most Exact Observations: Mapping Art and Science.* Saratoga Springs: Tang Teaching Museum and Art Gallery, 2001, pp. 80–81.
19. Denis Wood. 모든 것은 노래한다:이야기하는 지도들 (*Everything Sings: Maps for a Narrative Atlas*). Paju Book City: Propaganda Press: 2015.

Glas Journal: Deep Mappings of a Harbour or the Charting of Fragments, Traces and Possibilities

Silvia Loeffler

Abstract: With reference to a hybrid ethnographic project entitled *Glas Journal* (2014–2016), this article invites readers to reflect on the cultural mapping of spaces we intimately inhabit. Developed with the participation of local inhabitants of Dún Laoghaire Harbour, Ireland, *Glas Journal* seeks to explore the maritime environment as a liminal space, whereby the character of buildings and an area's economic implications determine our relationship to space as much as our daily spatial rhythms and feelings of safety. Deep mapping provides the methodological blueprint for *Glas Journal*. In order to create a heteroglossic narrative of place and belonging, I will contextualise the project with references to seminal works in the visual arts, literature, film and geography that emotionally map spaces. Chronotopes of the threshold will be used to elaborate on spatial and cultural phenomena that occur when crossings from public to private and interior to exterior take place. Touching upon questions such as "What is a space of protection?", "Who am I in it?", and "Who is the Other?", this article traces forms of liquid mapping that do not strive to conquer but rather to gain insight into the inner landscapes that are reflected in outer space.

Reprinted from *Humanities*. Cite as: Loeffler, S. *Glas Journal*: Deep Mappings of a Harbour or the Charting of Fragments, Traces and Possibilities. *Humanities* **2015**, *4*, 457–475.

Readings of spaces and places in terms of emotion go hand in hand with the creation of maps that function like a visual memory-cloth and concern issues like the personal, the underground, belonging, and longing. Here emotional mapping extends itself by going deeper, to become, indeed, deep mapping, and, in that sense, map-making becomes world-making.

This article introduces *Glas Journal: A Deep Mapping of Dún Laoghaire Harbour* (2014–2016), an ongoing deep mapping project of an Irish maritime environment that engages with its inhabitants' intimate routines and their attachments to the artificially built harbour landscape of Dún Laoghaire. Funded by the Irish Research Council as a postdoctoral project, *Glas Journal* hopes to reveal "unexpected" biographies of a specific harbour environment by investigating how the residents and workers of the given spaces between two piers relate to the sea as a border space and to the harbour as a space of protection. It is a visio-cultural cartography and will here be integrated into a broader examination of artistic forms of mapping that take place in the visual arts, literature and film, and which explore ambivalent affections, such as love and hate relationships, moments of transition, historical shifts, and the emergence of new identities in a spatial context.

Mixed ethnographic and artistic research methods are used to understand how a harbour is a shared space for a diverse set of inhabitants. *Glas Journal* involves individual and group participatory workshops that will be held at particular locations. The participants are invited to create site-specific hand-made journals about their living and work experiences, their everyday routines and habitats, and emotional attachments to places. The multiplicity of these stories, the particularities of place-based attachments, and the common themes emerging from the participatory workshops, will reveal a "place ballet" [1] based on people's feelings of attachment to "their" harbour.

To emphasise notions of belonging that may be found in a maritime environment and that support the perception of the harbour as a liminal, in-between space, a scenographic installation of Dún Laoghaire's two harbour arms, the West Pier and the East Pier, leading from the open sea to land will be set up in the former Mariners' Church, which now houses the Maritime Museum of Ireland. Site-specific books, which stand for various harbour locations along the seafront and are almost lined up like rosary pearls, as suggested in Figure 1, will then be put onto the representational harbour structure as if they are reference points on a map.

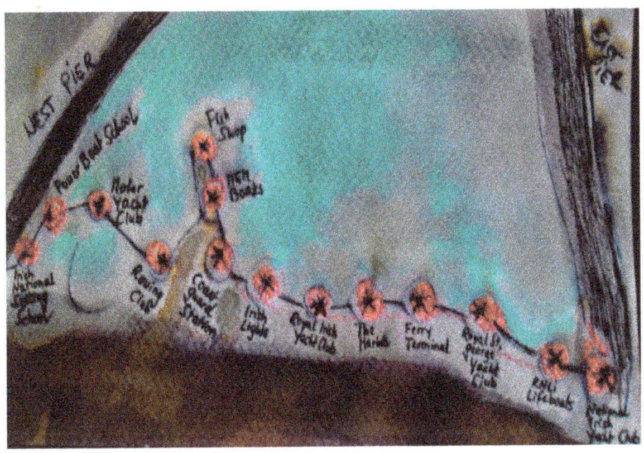

Figure 1. Map fragment, 2015. © Silvia Loeffler, Glas Journal, 2015.

1. Deep Mapping the Harbour as a Liminal, Spatial-Social Body

Dún Laoghaire, a small seaside town situated 12 kilometres outside Dublin, as a geographical place underwent dramatic changes in its development from the tiny 18th century fishing village of Dunlary or Dunleary to the imperial Kingstown of 1821 to 1920, which boasted a royal harbour and a magnificent railway line. The town's name after Irish Independence reverted back to Dún Laoghaire, and it became a thriving ferry hub in the 1990s. With the aviation industry already having taken over the leading role in global passenger transport, Dún Laoghaire, like the rest of Ireland, was hit hard by the economic crash in 2008, and when the Stena Line announced the final closing of its ferry service from Dún Laoghaire to Holyhead in February 2015, which had been operating since 1835, it felt like yet another ending with no new beginning.

The ferry boat model of the HSS Stena Explorer depicted in Figure 2 was donated by the Dún Laoghaire Harbour Company to the Maritime Museum of Ireland, where it has been restored and is to be part of the museum display. The particular image features as a fragment for the Ferry Terminal book in the *Glas Journal* series—because of the boat's obvious historical connection to the operational procedures in the harbour, but also because its containment in a glass vitrine mirrors the frozen quality of melancholia, where, in the attempt of evading psychic pain, without containment, there would be leaking.

Figure 2. Ferry Terminal fragment, 2015. © Silvia Loeffler, Glas Journal, 2015.

Current trends of rejuvenation, as part of a master plan to tackle Dún Laoghaire's ghost town status, include significant investment into cruise line business, the planned building of a Berlin-inspired urban pool/beach and a diaspora centre to attract visitors. The plans have been perceived as problematic in scale and validity to be of local benefit, and have consequently been contested by voluntary citizen initiatives and community organisations like Save Our Seafront (SOS) [2]. *Glas Journal* hopes to reflect these recent developments in the sense that residents and workers may evoke life cycles of birth, decay, destruction and resurrection, distinct from Dublin's phases of investment and its recent cultural and financial crisis. In addition to this attention to the socio-economic context, the project also reflects the physical landscape of the two harbour piers forming a space of protection, as the physical structure of the two piers (West and East) may be described as two arms that reach out to sea to guide naval vessels into the safety of land, as depicted in Figure 3.

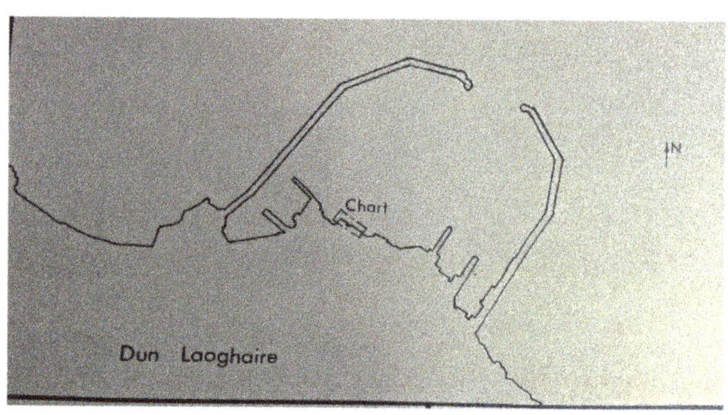

Figure 3. Map fragment, 2015 [1983]. © Silvia Loeffler, Glas Journal, 2015.

The original map fragment showing Dún Laoghaire Harbour in Figure 3 is property of the Dún Laoghaire Harbour Company, who kindly allowed me to photograph their "old" maps in storage. The elegant minimalist fragment originally served as a hydrographic map and also has a stamp from the engineering office of the Commissioners of Irish Lights on it, which dates back to 8 June 1983 (the 1980s being another era of economic depression and mass emigration in Ireland). In 2015, the fragment stands for an almost pathological desire for the map to be reconstructed or to be coloured in.

The concept of liminality thus applies to times of transition, historical shifts and the emergence of new social and cultural paradigms. To cross the threshold into a new aeon, spaces themselves are required to shed their skin and to adapt a new identity. Liminality as a concept of cultural hybridity often refers to the disruptive nature of inbetween-ness and the interstitial pathways or passages that constitute the body of territorial maps and their spatial divisions. The threshold space in-between boundaries was called *limen* or threshold by the Belgian anthropologist Arnold van Gennep in *Rites of Passage* (1909). Liminality as a concept has been used by Mircea Eliade [3] to divide the human experience of what is sacred and what is profane, whereas liminal space was regarded as transformative by Victor Turner, the ritualistic passageway from one state to another [4].

If we consider these various conceptualisations of the threshold or liminal space in relation to harbours, we understand that they are not just nodes in a network. Rather they are places and localities that emerge from numerous threshold crossings to recount the diverse social lives of the harbour as border space providing a rich source of information about the rituals *and* buildings, homes *and* docks, myths *and* quantitative measures of the sea. Harbours are often seen as sites of recreation (boat docking, fishing, sailing or waterside redevelopment) or as export nodes of a global economy. However, in Allan Sekula's photographic ethnography, *Fish Story* (1995), harbours are not havens any longer because the sea now equals money [5]. In support of Sekula's findings about the effects of an excessive capitalist economy on the sea-faring industries, Michael Taussig elaborated on how ports have become gentrified, how museums contain the objects that were once vital for survival, and how luxury housing with a "sea view" replaced working class areas. But when describing his personal relationship to waves, Taussig also mentions a "fragile freedom, always already lost" ([6], p. 260) which brings us to the magic of liminal spaces inbetween commodified agendas and to the multiple meanings of threshold, where we pass from one phase to another in terms of ritual and initiation, historically and personally, as well as spatially.

Mikhail Bakhtin's literary concept of the chronotope [7], which refers to the cross-section of time and space in the novel that determines the plot, may be used to map historical and geographical changes that bring about the various in-between states that building structures find themselves in. The relevance of mapping in-between or liminal spaces lies in the exploration of psychological meanings of spatial attachment to and/or detachment from places of home and work. A cyclical theme that appears to emerge again and again in the context of home, exile and belonging is how tightly interlinked mappings of home are with feelings of safety within the community and the role that spaces of protection play in everyday life. Inspired by Walter Benjamin's staccato methods of mapping the everyday that culminated in the fragmented form of a collage or montage of cultural meanings [8], *Glas Journal* seeks to explore the emotional dimensions of spatial attachment to a harbour, which, by its nature, is a space of protection.

Drawing upon a methodology then that is based on a visio-textual bricolage consisting of fragments to survey forms of life, as Benjamin would have it, *Glas Journal* looks at how the themes of care and intimacy are interwoven in a complex contained public space such as the harbour. By seeking to emphasise how individual stories, performances, and emotional geographies are intertwined with the physicality of a built maritime environment, the first series of 14 hand-size and hand-made artist books that document 14 corresponding harbour spaces is created by me in my studio in the Coast Guard Cottages. This process of making allows for a positioning of the subject matter of deep maps in the fragmented form of artist books as well as for a positioning of myself as the initiator and participant of the project. The second series of books will be based on 14 journals made in collaboration with the residents and workers of the given spaces between West and East Pier, in order to map how they relate to the sea as a border space and what the harbour as a space of protection means to them. A form of emotional mapping that deals with the harbour as an allegory creates a cartography of intimacy that serves as an antidote to the sharp distinctions of borders and separation lines as used in the classified western mapping system to define territories.

2. The Modern Map as Icon of Territorial Space

Maps in their conventional appearance are two-dimensional visual representations that aspire to help the user to get from one place to another in a geographical context. Their symbolic contents allow for an understanding of place, they mirror social and political decisions. Maps express a point of view. They indicate property lines, postal districts and enterprise zones, and they are always biased in the sense that they project the interests of their creators, as Wood states in *The Power of Maps* [9].

Having moved from the imaginary landscape representations found in the cave drawings, stone engravings, clay and papyrus maps of the ancient world, via the visionary geographical compositions of monastic scholars and the lavishly beautiful medieval *portolani* of sailor-mapmakers, the modern map in its initial form, in enlightened scientific discourse, constituted a "world picture" that was based on mathematical precision, where "geometrically defined spatial coordinates" were seen as "the guarantor of [a map's] neutrality". The grid-like rational map came with the claim to be free from the constraints of ideological factors like sex, politics and race and was seen as an unbiased objective and accurate tool to direct its viewers from one place to another ([10], p. 148).

As opposed to being fluid and performative, this scientifically sound form of mapping created a unified understanding of space on fixed terms. In *Imagined Communities*, Benedict Anderson regards the mechanically reproduced European-style map as a fundamental building block in the development of the idea of the nation-state. Western maps derived from a cartographic model that expressed ownership and control, and they were used as instruments of classification in the attempt to administer the state more efficiently. Maps by their nature are influenced by the impressions of their creators and reflect a certain type of reality. In a colonial discourse they were seen as the representation of a body of knowledge, a pictorial demonstration of power. Anderson states that the construction of colonial thinking was reflected in the interlinking of the map, the census, and the museum as mechanisms of control and neat classification, whereby "comic classificatory and subclassificatory census boxes entitled 'Other' concealed all real-life anomalies" ([11], p. 184).

If geographical space in the modern era was primarily defined by forms of mapping that were based upon the establishment of borderlines, a discussion about being outside those lines, or about being "in-between" arises, which leads to an evaluation of modes of belonging and an emotional reality that could not exist in the absence of maps. Cartographies that reflect a cultural and emotional discourse of space form a complete contrast to maps that reflect nationalist imaginings of power deriving from occupied and colonised territory. One of the central themes in the work of British-Lebanese artist, Mona Hatoum, deals with elements of diaspora and considers how identity is shaped by absent feelings of belonging. *Map* (1999), an installation of identical glass marbles forming the world's continents on the floor, challenges the sharp distinctions of borders and separation lines used in the classified western mapping system. When the viewers of Hatoum's *Map* walked the parquet floor, the marbles started moving. It was possible to slip on those glass balls, and in this sense unstable ground was created, challenging the viewers' perception of belonging and evoking the experience of liminal exile space as a wobbly territory of contradictions. About a decade later, another one of Hatoum's maps shows the continents in moth-eaten fashion engraved into a carpet (*Afghan [red and black]*, 2008), still questioning the reliability of a map and the security of home [12].

3. Tender and Chaotic Mappings

The field of emotional mapping offers a new form of power to maps—it is an engagement of spatial practices with the human condition, and, in that sense, it seeks to redeem the calculating and colonising tactics of forms of cartography that feed into the fantasy of territorial sovereignty. The focus on emotional responses to space, how one interacts with the public sphere in private and intimate ways, shifts the investigation of the city as a web of streets defined by their architectural setting into a search for possible alternative interpretations of what it might mean to navigate through a city visually and emotionally.

Numerous important aspects of emotional mapping can be grasped in the poem "Your Street Again" by Sophie Hannah [13]:

"Guess who I saw last night', was all she said.
That, and the answer (you), was all it took,
And now I'm leafing through my A–Z
To find your street again. I had to look
Four years ago, and memorize the way:
Palatine, Central, Burton—halfway there.

I don't intend to visit you today
As I did then, and so I shouldn't care
Which road comes after which and where they lead.

I do though. I repeat them name by name
My house is here and yours there. I need
To prove the space between them stays the same."

A connection between places, signs and emotions has been established, a connection that has been hinted at, and thoughts that may fleetingly cross our mind—yes we know what it means to look for someone's street, we can relate to that—although it is rather difficult if not impossible to express what Sophie Hannah *really* meant by "Palatine, Central, Burton—halfway there". What does her mapping stand for? Is it an allegory for a doomed relationship, someone departed she does not want to cross paths with ever again, a spatial mind exercise of how long it takes from Palatine to Central to Burton? Hannah's poem evokes a deeply intimate process of mapping, something so important in her spatial memory that, in order to get it right, she needs to get her A–Z (because someone has been seen, so he or she is still visible and creates an impact) and "Palatine, Central,

Burton—halfway there" act as her emblems to make sense of the situation ("I need to prove the space between them stays the same").

The Situationist International movement operating in Paris in the 1950s and 1960s created collage-based maps representing psychological influences and playful associations with space. The Situationists' maps asked to be examined in their function as a resistance to space that was determined by the ideology of a hegemonic state apparatus, with its focus on capitalist market values and consumer-orientated forms of representation. They called for an exploration of the concept of "public intimacy", which relates to the entanglement of personalised expressions in public space and their possible meanings. One of the major influences for the Situationists was Madame de Scudéry and her tender map about intimacy and the discourse of love, which goes back to the 17th century. In 1654, Madeleine de Scudéry published her novel *Clélie*. She designed a *Carte de Tendre* or Map of Tenderness and created an emotional topography with roads called Friendship to cities with names like Tenderness, Great Wit or, negatively, Submission—all concerning the affairs of the heart, with the hope of ending up in The City of New Friendship. At first hearing this may not sound extraordinary, but de Scudéry's use of an exterior that expresses an interior landscape opens up new approaches to the mapping of space ([14], p. 264).

De Scudéry's form of mapping may be compared to the workings of cathedral space, where an inner sanctum is created through gothic shape, the sound of music and the smell of candle wax, transporting the praying person into an inner refuge [15]; or, on the opposite end of the emotional scale, to the works of Rainer Maria Rilke, who consistently wrote about the chaos of inner space translated into the terminology of an outer world in his works. In Rilke's only novel, *The Notebooks of Malte Laurids Brigge* (1910), the protagonist is not aiming at a detailed description of modern life in metropolitan Paris but isolates certain places and injects them with personal meaning [16]. Malte is not concerned about depictions of urban "realities" but rather projects his own inner psychological space into the outside of the city. His experiences with urban decay and his encounters with the city-dwellers and their fear of poverty and death result in moments of paranoia for Malte. Malte's wanderings through Paris are a form of becoming. The city becomes his emotional states and his emotional states become the city.

Another example for an emotional mapping of the city is Wim Wenders' *Wings of Desire* (1987), which refers to Berlin's post-war void. Dealing with spatial erasures and the despair of modern city life, this film can be seen as "an architectural document of a city that no longer exists" [17]. The work's moving images dissolve the frozen quality of melancholy by paying tribute to discoloured urban space with care and empathy. Wender's inserts a poetic discourse into everyday language by deploying an angelic phantasmagoria as a perspective that allows one to view the modern city in psychological terms. In the film, urban angels, that yearn to become human, offer a perspective from above: an angelic perception onto the everyday activities of humans. The aerial perspective of the city is offered to the viewer and this contributes to a voyeuristic feeling of witnessing the intimacy of people's lives and by being provided with a bird's eye view the viewer looks into people's apartments, into their thoughts. *Himmel über Berlin* (the film's original title) can be considered a cinematic memory cloth of 1980s Berlin because of its interlinking of personal stories with the topography of the city, black and white and bleak in representation changing into colour, once the protagonist of the film, an angel, turns human and is able to live the city experience of the here-and-now, the city to becoming a space that "could be characterised by the experience of procession and movement, of appearance and spectacle, of progress and the new, of dreams and ghosts" [18].

Opposed to the Situationists' disillusion with modern urban life expressed in Guy Debord's final statements that *logos* had triumphed over *eros* by turning the city into nothing more than cheap-thrill commodity zones, Wenders' film is leading to the act of "logos redeeming eros" [19]. In that sense phantasmagoria as an inherently fluid concept was used to map the threshold experiences of a city in transition, whereby its liminality was reflected back and forth in the corporeal metamorphosis of the film's protagonist as well as in the film's layered narrative of belonging and longing. This kind of layered or liquid mapping, whether it is expressed in tender ways in de Scudéry fashion or more chaotic and dreamlike, is the essence of *Glas Journal*, where 14 artist books covered and stained by ink layers reflect the colour of the Irish Sea, as illustrated in Figure 4, and serve as the basis for a contemporary map of shelter and protection, of navigation and connection, and of processes of anchoring and homecoming. The books are located in a refuge inbetween two scenographic harbour arms, but the exhibition venue (the former Mariners' Church now housing the National Maritime Museum of Ireland, which is about ten minutes walking distance from the East Pier of Dún Laoghaire Harbour) in itself may be described as a space of guidance, and, possibly, healing.

Figure 4. *Glas Journal* installation, Maritime Museum of Ireland, 2015. © Silvia Loeffler, Glas Journal, 2015.

4. Cartographies of Intimacy

In terms of gendered place-making, the issue arises as to how emotional mapping and a form of writing that does not exclude women and their spaces can be created. If the social functions of artistic production and presentation are being viewed from a feminist viewpoint, past history has rarely included women artists. Geography as well as fiction is a discourse that has been shaped by male dominance for a long time and created disciplines that Rosalyn Deutsche calls a "boys town" [20]. In *Atlas of Emotion* (2002), Bruno extends Virginia Woolf's concept that a woman must be able to call a room her own and to have some money in order to write fiction to processes of emotional mapping. Bruno refers to Gertrude Stein's literary-geographic path, where geography stands for a way of writing that Bruno calls "tender cartography". Spatial representations that include women in the field of geography become a "room of one's own" ([14], p. 209).

The use of elements of oral history in *Glas Journal* to map difficult threshold experiences in a handmade book is fundamentally different to the traditional western cartography from the 17th century onwards, which was based on spatial representations that depicted the development of European nation-states and defined territorial claims of ownership. When Mei-Po Kwan used geographical information systems (GIS) to map the levels of fear that are prevalent in the everyday lives and quotidian routes of Muslim women in the US post-September 11, she challenged the concept of a binary us-versus-them opposition that is conveyed by single-voiced authoritarian narratives [21]. In her study of 2008, Kwan pointed out the importance of oral history accounts, where participants tell about their threshold experiences in terms of their memories and emotional connections to events, people and places. In order to construct her geo-visualised multi-media narrative of the life paths of a Muslim woman based in Columbus with the fictitious name "Nada", Kwan used daily activity data and emotional coding to identify "Nada"s levels of fear. Kwan made the important point that women, who have been made *Other* through a dominant anti-Muslim narrative by being associated with the category of "terrorists and anti-American foreigners" ([20], pp. 666–67), suffered from significant fear on their daily routes and in public places in particular, and that there needs to be a shift of perspective as to how these fears are being constructed. Kwan also asked for a more careful examination of the life paths of Muslim women in the US, rather than creating yet another homogenous category for their experiences.

In questioning traditional cartographic models and their alleged objectivity, an aesthetic of failure of how we read representational as well as experiential maps may be necessary, in order to enter different paradigms of thinking. Juxtaposing, or, perhaps, complimenting Kwan's representation of post-September 11 experiences of Muslim women in the US, Iranian born artist Shirin Neshat's oeuvre, in particular her black-and-white portrait series of *Women of Allah* from the 1990s, is based upon an orchestrated rebellious fearlessness within dogmatic social and cultural systems that aim at keeping women contained in the domestic sphere. Neshat, who has been living in American exile since the mid-1970s, uses several layers to construct her fantasy of the Muslim woman as a revolutionary. In Neshat's early photographs, "woman" always wears a chador, often points a gun, its barrel sometimes appears to be in the place of earrings. And then there are the layers of Persian calligraphy. At first glance the Farsi poems of contemporary Iranian women appear to be neatly inked on to the face of "woman", her hands or her feet. On closer inspection the viewer realises that the writings are meticulously applied to the prints, rather than directly tattooed on to exposed skin. The inky calligraphy writings are confusing mappings that create barriers for non-Farsi speaking viewers, who may be forced into perceiving these texts in their graphic beauty as mere decoration, or as cultural codes that are incomprehensible [22], beyond grasp of their meanings, and hence experience feelings of failure and a longing "to understand the map".

Glas Journal is an organically growing multi-layered project that allows the participants and myself to deal with writing as a source of cure and poison simultaneously, in order to come to terms with moments of crisis, whether economic, personal or spiritual, and to overcome the binary oppositions of remembering/forgetting. Deep Mapping processes relate to unstable states and ambivalent emotions that may lead to the discovery of fragments that are confusing. However, as deep maps allow for multiple expressions and interpretations, they cannot "go wrong".

5. Liquid-Coloured Mappings of Belonging and Longing

In Irish, the word *glas* is reserved for the indefinite shades of green, blue and grey that are present in the sea. I take this chromatic generosity as a marker for *Glas Journal*. By using layered colour occurrences as a metaphor for the multiple processes that occur in works of deep mapping framed by artistic narratives, I seek to transform binary oppositions of spatial laws and their fixed regularity into a fluid and performative discourse. A "translation" of colour semantics taking place, as suggested in Figure 5, where *glas* as a colour of nature and the Irish Sea in particular becomes a visio-cultural register for cultural liquidity in *Glas Journal*.

Figure 5. *Glas* coloured Irish Sea and *glas* on paper, 2014. © Silvia Loeffler, Glas Journal, 2015.

Glas as a colour in old Irish and in Welsh is also used for grassy patches and the skin tone of dead animals, both rather confined, opposed to the general idea of "the boundless, infinite sea" [23], but in its liquid function it could not be applied to the Pacific, for instance, or to the Mediterranean Sea, which Homer described as "winedark" in *The Illiad*.

> Homer's sea, whether *háls*, *thálassa*, or *póntos*, is described as misty, darkly troubled, black-dark, and grayish, as well as bright, deep, clashing, tumultuous, murmuring, and tempestuous—but it is never blue. The Greek word for blue, *kuáneos*, was not used of the sea until the late sixth or early fifth century BC, in a poem by the lyric poet Simonides—and even here, it is unclear if "blue" is strictly meant, and not, again, "dark" [24].

The Roman word *locus* (deriving from "to stand") equates to the Greek *topos*. The ancient Greeks used *topos* for any kind of place, if it were to be found in a book, body, landscape or any other kind of discourse. *Chora*, the Greek used for larger settlements, relates to the verbal description of places. In Hellenistic Greece, chorography referred to the quality of place, to the description of a given area and all its particularities. Traditional cartographers, like surveyors, aim at an exact visual reproduction of space, whereas chorographers create a medley of geography, topography and history of a place that is based on a heteroglossic narrative. Joseph Rykwert describes chorophilia as an archaic expression that may be used to describe the relationship between an exile and their homeland, and states that Joyce's *Ulysses* is "one of the grandest monuments to the non-returning exile"s chorophiliac nostalgia' ([25], pp. 14–16). Contrasting colour references of the sea are found in numerous literary accounts of aquatic locations. A sense of familiarity, the taste and smell of a place, its climate and its colours, contribute to emotional stability and security. If one is deprived of this sense, nostalgia comes into being. How do personal ties to specific locations come to be understood as signifiers that encompass the human desire to feel safe in landscapes that were formerly shaped by high productivity, but now appear to have been taken over by notions of emptiness, melancholia and ghosts from the past?

Whereas Homer talked about the "winedark" sea in a Greek context, Joyce did neither use "glas" nor "winedark" but "snot green" for the Irish Sea. Joyce's reference in *Ulysses*, where he describes "the mailboat clearing the harbour mouth of Kingstown" (Dún Laoghaire's name when it was part of the British Empire) brings up once more the fact that places in transition phases often change colour as well as names.

> He mounted to the parapet again and gazed out over Dublin bay, his fair oakpale hair stirring slightly.
>
> God, he said quietly. Isn't the sea what Algy calls it: a great, sweet mother? The snot green sea. The scrotum tightening sea. *Epi oinopa ponton.* Ah Daedalus, the Greeks. I must teach you. You should read them in the original. *Thalatta! Thalatta!* She is our great, sweet mother. Come and look.
>
> Stephen stood up and went over to the parapet. Leaning on it he looked down on the water and on the mailboat clearing the harbour mouth of Kingstown [26].

The original map fragment of Kingstown shown in Figure 6 is property of the Dún Laoghaire Harbour Company and currently in storage in the Ferry Terminal. It is a beautifully drawn drainage map that depicts a winedark border line marking a "proposed intercepting sewer" and dates back to 29 September 1873, a year when Joyce was not even born yet, and Kingstown/Dún Laoghaire basked in the afterglow of a building boom, having ten years previously been described as "the wealthiest and most popular township in Ireland" ([27], p. 99).

The use of a dark red borderline leads to a contemporary interpretation of Homer's use of "winedark" as an aquatic colour and brings up notions of the blood red of water because of the death of thousands of migrants on maritime borders. The winedark sea ties in with passages across the water to reach the place one may call home, and to belong. Given the humanitarian disasters of boat people in the Mediterranean Sea and in South East Asia, the term "Asylum Harbour", which is what Dún Laoghaire Harbour was historically called before it became the "Royal Harbour of George IV at Kingstown" ([27], pp. 20–21, 26), takes on new meanings of large-scale global dimensions. Mapping the theme of belonging asks, perhaps more than anything, to put displacement on to a map, and to give it a name and a home in a space of protection.

Figure 6. Map fragment, 2015 [1873]. © Silvia Loeffler, Glas Journal, 2015.

"What is a space of protection?", "Who am I in it?", and "Who is the Other?" are main research questions in *Glas Journal* in order to establish a model of cultural practice that attempts to create a broader understanding of visual and spatial experiences on a local level in Dún Laoghaire Harbour. The enquiry does not focus on the mastery of a certain historio-geographical discourse, but stresses the importance of the bricolage, the fragmentation and the in-between in its production of meanings. One of the concerned harbour spaces that is being deep-mapped for *Glas Journal* and evokes feelings of stability and safety but has undergone profound changes on liquid ground is Dún Laoghaire's old coastguard station with the adjacent coastguard cottages. Renowned maritime historian John de Courcy Ireland stated that one of the functions of the British Coast Guard service operating in Ireland from the 1830s onwards until independence was to help vessels in distress, and that the service, which "was normally manned by former naval men and worked under semi-military discipline" [28], was strongly successful in saving lives, albeit the local population was inclined to support smugglers and opponents of British Rule.

Dún Laoghaire's Coast Guard Station, as shown in the map fragment of Figure 7, the original being stored by the Dún Laoghaire Harbour Company and dating back to 1911, a year where the township of Kingstown was still on the rise, echoes the boundary status and dialectic of the liquid threshold in its spatial representation.

Figure 7. Map fragment, 2015 [1911]. © Silvia Loeffler, Glas Journal, 2015.

Currently only three of the eight residential units are occupied. Two units are family homes, and one cottage is home to the Dún Laoghaire Arts Centre downstairs, and three individual artist studios upstairs (one taken over by myself). Some of the empty units are used as storage space, and, occasionally, as a film set.

All throughout Ireland there are similar styles to "my" Coast Guard Station. The watchtower stands tall over the residential units that were initially intended for crew members and their families. The layout itself is a reminder of Jeremy Bentham's *panopticon* used for institutional architecture, like the asylum, the hospital, and the prison. Yet I love this building, the fragments depicted in Figure 8 paying only small homage to its magic. I have been working in the Coast Guard Cottages for three years. I feel the building's solidity and peace, its proximity to the sea. Its severe structure is softened by the lush gardens and wildly changeable coastal weather. I often look out of my studio window over the gardens to the West Pier and into Dublin Port and wonder who has had that view before me.

Figure 8. Coast Guard Station fragments, 2015. © Silvia Loeffler, Glas Journal, 2015.

Allen, artist son of a sea captain, and his group of painters, the Dún Laoghaire Art Centre, have occupied several locations in the harbour throughout the years and have been based in the Coast Guard Cottages for almost a decade, first in Number One, and now in Number Three. Miriam, one of the fellow artists and a local of Dún Laoghaire, kept a communion certificate that she found at the time when the studios were based in Number One. The document conveys that a resident who lived in the Coast Guard Cottages approximately one hundred years ago, a little girl called Maisie McDonald, made her Holy Communion on 27 May 1916, one month after the Easter Rising, when an independent Irish state was established. As it happens, 27 May is also the birth date of Miriam's daughter, who is now the guardian of the fragment of Number One, Coast Guard Cottages, which is depicted in Figure 9.

Figure 9. Coast Guard Cottages fragment, 2015. © Silvia Loeffler, Glas Journal, 2015.

One of the central issues of linguistic anthropology is the importance of language and place names, the use of metaphor as well as the uses of silence, blank spaces and gaps, which help us to define our modes of belonging on liquid grounds, and *Glas Journal* hopes to open the doors to spaces we knew nothing of before we were handed a snippet of random information. When describing his study of Western Apache senses of place, Keith Basso points out that the ways in which native people communicate about their landscape "may shed light on matters other than geography" ([29], pp. 101–2). "Belonging" in the English language originally had spatial implications rather than referring to possession or ownership. Boullata compares this meaning to the word's Arabic roots. In Arabic "belonging" (*intima*) refers to an activity—to develop, to make progress, and its original English meaning refers to a relationship between two things that "may be equally long and run parallel", they share something significant, require each other. The Yanomami, a semi-nomadic tribe in the Amazon, on the border between Brazil and Venezuela, perceive themselves as belonging to the land, as opposed to the land belonging to them. They consider their territory as alive, in constant change and movement and do not live with the concept of the border as a line that is not to be crossed [30].

When contemporary Islamic artist, Emily Jacir, an American-Palestinian, who is able to travel freely where displaced Palestinians are not allowed to go, collected people's wishes that were mostly of a mundane nature, she went about to look for somebody's house, visit somebody's mother's grave, and pay a phone bill for a man in the post office of Jerusalem. Jacir followed in Walter Benjamin's footsteps to trace life, in her case, the vanished life of others, to negotiate meanings of home. Whereas Mona Hatoum's maps and spheres often relate to objects of domesticity (like colanders, graters, soap bars and neon tubes), and where the objects extend themselves to being unstable maps because they are objects without belonging to a place, Jacir's approach to people who live in exile was to weave their fragments of meanings of home into a metaphorical tapestry of connections. She acted on behalf of the exiled person's desires and travelled to Jerusalem or Haifa to carry out their wishes, brought very mundane things back and forth, or attended events on their behalf, and as such gave a narrative to the history of displaced individuals whose former home in its physicality is unattainable to them. By collecting these fragments to document *Where We Come From*, 2001–2003, which first appeared as a magazine publication on Palestinian territory [31], Jacir's work deals with mapping others' (and perhaps her own) up-rootedness. When Jacir crossed geographical borders for those who are not allowed to do so freely, she also crossed boundaries between arts practice, individual life histories and politics by producing a cultural work that emphasises the multi-faceted textures of spatiality, travelling and belonging.

In her work *Atlas of Emotion* (2002), Giuliana Bruno insists on the entwined nature of seeing and travelling. She connects "emotion" with "motion", "sight" with "site" and metaphorically draws polyphonic interdisciplinary maps that are based on cultural life. In accordance with Bruno's work, a creative form of mapping that gives information about places and spaces in a visual context, and that differs from traditional cartographic and architectural models, was chosen as the topic of enquiry for *Glas Journal*. To elaborate upon mapping techniques that defy the box-ticking systems of control of classified western cartography, and to integrate forms of belonging and displacement on a deep map, also means to integrate forms of longing. The hand-made artist books in *Glas Journal* that are currently in the process of being made, as demonstrated in Figure 10, stand for individual places in the harbour and are included as a valuable source of information for the visual emotional vocabulary that constitutes urban palimpsests (the palimpsest referring to the multiple possible readings of public space as a text), wherein arts research practice criss-crosses with cultural geography, visual culture and critical theory.

Figure 10. Glas Journal fragments, 2015. © Silvia Loeffler, Glas Journal, 2015.

The idea to put a spatial discourse into books goes back to the idea of inner space, of finding a way to express an inner infrastructure that captures modes of being. The books reveal themselves as sites of emotional habitat, bearing to stay with the spaces of the present that differ from a colonial past, but still carry its heritage. A liquid or deep way of mapping may expose some aspects of this patrimony, but has no interest in "fixing" history. Rather, the attempt is being made to relate "the story" or "the image" to pictures, scenes and sounds of everyday life and to bring a fragmented discourse, where dispersal and disintegration are woven into places, home.

6. Conclusions: Journeys and Journals as Wayfinders

Multi-sited or hybrid explorations of ethnographic space asks for new cultural productions of spatial meanings. In order to create such an alternative cultural production that concerns the public sphere of the harbour with a narrative that is built upon a multiplicity of different voices, I am looking at places "through discovered metaphorical associations", as Marcus terms it ([32], p. 108). The artist books in *Glas Journal* stand not only for places but also for representational forms of feelings that constitute a fluid and multi-sensory map of the harbour. I chose for a hybrid assemblage of narrative structures because the patching together of a varied amount of cultural material allows for an exploration of intimacy with public space in a way that I could not have grasped with the objectively structured methods of a scientific analysis, resulting in a map that is "to be trusted".

As shown in Emily Jacir's working to produce memory cloth of Palestine for exiles, where she interweaved spatial attachment strands of individuals to a homeland that only exists in memory, an emotional mapping of spaces that includes wishes and desires cannot be easily understood through questionnaires or public opinion surveys. Instead, fragments from personal narratives provide insights about difficult feelings that circuit around loss, change and insecurity. Gaps, errors and discontinuities of an emotional discourse will also offer significant "data" about the atmosphere of a place, including its mood and energy. From these materials, a map can be created that situates local knowledge through the space-times of cultural myths, local icons, and shared histories.

The hybrid forms of how we make sense of places are muddled, contrary and in-between by its nature, and prove difficult to be pressed into a specific typology of rigid parameters. If experiential explorations of being in a space, and of subsequently creating a journal that functions like a memory trail by extracting personalised symbols of oral history accounts are dealt with as patchwork pieces, the fragmented narrative allows for a circular-structured investigation and randomness that addresses the ambiguity of threshold spaces. This kind of approach leaves it up to "the participants", "the creators", "the travellers of their own spaces" to inject their journals with cultural meaning. It also functions as a tool to critique the linear concept of borderlines that divide A from B, and, in this way, to simultaneously question the satisfaction that the Social Sciences derive from finite and neat chronological narration models.

Writing enables self-reflection, but by putting the Self on to a map that is always unfinished, incomplete and contradictory, the writer has to face up to the ambivalence of intimacy, in order to reveal what their day may literally and visually be like. Sometimes, it is not possible to put feelings or memories on a map, particularly if meant for public display, and so the page stays empty, and, perhaps, left to the viewer's imagination, like in Figure 11.

Deep mapping is the process of collecting fragments in order to gain understanding of liminal spaces. In the creation of *Glas Journal*, the participants and myself make artist books that contain map segments, drawings, paintings, photographs, sound recordings, and objects, showing puzzling connections and intersections, exciting loose bits and frustrating gaps that form a different kind of topography—a deep map. My mapping of Dún Laoghaire Harbour hopes to make visible a larger story about place, belonging and public intimacy, contributing to interdisciplinary debates where space is more than a "flat surface" that we travel across. It co-constitutes the social through the simultaneous, unfolding pathways of life stories ([33], p. 5). Deep Mapping is very often about rituals of space. Iain Biggs introduces Anthony Lyons' deep mapping project of the Quantock Hills in Britain, which was called *Quantock Dreaming. Secret Mappings and Mapping Secrets* (2008), and which he conducted with members of the public and in co-operation with poet Ralph Hoyte, as a "slow residency" [34]. The term resonates with Werner Hamacher's philological treatise and his notions of waiting. Before the expectations waiting was there all along, which in German is a lovely word play: "Vor der Erwartung war das Warten" [35]. Waiting for Hamacher appears to be the space where one is holding back as well as holding open. The space of waiting is full of hidden meanings, and it is liminal.

Figure 11. *My day* fragment, 2015. © Silvia Loeffler, Glas Journal, 2015.

Glas Journal reflects meanings of the sea, of home, of harbour. The deep mapping work for *Glas Journal* in the form of a slow residency intertwined with waiting happens in the Coast Guard Cottages (the studio site), on the harbour stretch between West and East Pier (the work site), in the Maritime Museum (the archival site), and—literally—in the Irish Sea (the swimming site). In connecting all these sites spatially, physically and artistically on a daily basis, several crossings take place that relate to historical and individual passages, almost pilgrim-like constructs of journeys at times. By looking at the community function of "containing" wholesome internal space, as opposed to the fragmented and uncontained, which is accompanied by the melancholia of loss and nostalgia for a Golden Age, the harbour as a womb-like space may also be interpreted as a sacred space that has the translucent quality of water leading from darkness to light.

The creation of an artistic harbour installation as a space of protection with artist books that form the reference points of maps speaks of the possibility to accommodate multiple narratives about meanings of liquidity as well as of solidity and shelter. Even though all structures are unstable, public intimacy as a concept for the empathetic exploration of spaces is very exciting because it relates to a powerful public engagement concerning the production of space by looking at its performative elements. Deep mapping is also about the re-claiming of ownership of living and work places by creating a many-faceted framework of spatial understanding and weaving it into a rich multi-sensual narrative by posing the crucial question of "How do you feel connected to this place?"

Conflicts of Interest

The author declares no conflict of interest.

References and Notes

1. David Seamon. "Body-Subject, Time-Space Routines, and Place-Ballets." In *The Human Experience of Space and Place*. Edited by Anne Buttimer and David Seamon. New York: St. Martin's Press, 1980, pp. 148–65.
2. SOS/anon. "The Future of Dún Laoghaire Harbour & Proposal for Super Cruise Berth." Available online: http://www.saveourseafront.net/ (accessed on 17 August 2015).
3. Mircea Eliade. *The Sacred and the Profane*. Orlando: Harcourt Publishers, 1987.
4. Victor Turner. *The Ritual Process: Structure and Anti-Structure*. New York: de Gruyter, 1995, p. 94.
5. Allan Sekula. *Fish Story*. Düsseldorf: Richter Verlag, 1995.
6. Michael Taussig. "The Beach (A Fantasy)." *Critical Inquiry* 26 (2000): 248–78.
7. Mikhail Bakhtin. "Form of Time and Chronotope in the Novel." In *The Dialogic Imagination: Four Essays*. Edited by Michael Holquist. Austin: University of Texas Press, 1981, pp. 84–258.
8. Davis S. Ferris. *Walter Benjamin: Theoretical Questions*. Stanford: University Press, 1996.
9. Denis Wood. *The Power of Maps*. New York: The Guilford Press, 1992, p. 24.
10. David L. Martin. "Cartographic Desire: Mapping and the Erasure of Space". In *Curious Visions of Modernity*. Cambridge: MIT Press, 2011, pp. 147–76.
11. Benedict Anderson. *Imagined Communities*. London: Verso, 1991, pp. 181–85.
12. Kirsty Bell. "A Mapping of Mona Hatoum". In *Unhomely*. Berlin: Holzwarth Publications, 2009, pp. 61–73.
13. Sophie Hannah. *The Hero and the Girl Next Door*. Manchester: Carcanet Publishers, 1995, p. 40.
14. Giuliana Bruno. *Atlas of Emotion*. New York: Verso, 2002.
15. See Tuan's comparisons of the modern skyscraper to the Gothic cathedral in Yi-Fu Tuan. "Time, Space and Architecture. Some Philosophical Musings." In *Topophilia and Topophobia. Reflections on Twentieth Century Human Habitat*. Edited by Xing Ruan and Paul Hogben. London: Routledge, 2007, pp. 22–30.
16. Rainer Maria Rilke. *Die Aufzeichnungen des Malte Laurids Brigge*. Berlin: Insel Verlag, [1910] 1982.
17. Robert Phillip Kolker, and Peter Beiken. *The Films of Wim Wenders. Cinema as Vision and Desire*. Cambridge: University Press, 1993, p. 15.
18. Steve Pile. *Real Cities: Modernity, Space and the Phantasmagorias of City Life*. London: Sage, 2005, pp. 20–21.
19. The Situationist International protested against the functionalist rationale of modernist architecture: "Like Henri Lefebvre, they [the Situationists] concurred that in the modern city Logos had triumphed over Eros, order over disorder, organization men over rebels." See Andy Merrifield. *Metromarxism: A Marxist Tale of the City*. New York and London: Routledge, 2004, p. 95.

20. Rosalyn Deutsche. "Boys Town." *Environment and Planning D: Society and Space* 9 (1991): 5–30.
21. Mei-Po Kwan. "From Oral Histories to Visual Narratives: Re-Presenting the Post-September 11 Experiences of the Muslim Women in the USA." *Social and Cultural Geography* 9 (2008): 653–69.
22. Valentina Vitali. "Reading Mahdokt." In *Shirin Neshat*. Edited by Britta Schmitz and Beatrice E. Stammer. Berlin: Nationalgallerie im Hamburger Bahnhof, 2005, pp. 107–11.
23. For a more nuanced discussion of shades of *glas*, see Máiréad Nic Craith. *Narratives of Place, Belonging and Language: An Intercultural Perspective*. London: Palgrave Macmillan, 2012.
24. Caroline Alexander. "A Winelike Sea". *Lapham's Quartely*. Available online: http://www.laphamsquarterly.org/sea/winelike-sea (accessed on 31 May 2015).
25. Joseph Rykwert. "Topo-philia and -phobia' Identity." In *Topophilia and Topophobia. Reflections on Twentieth Century Human Habitat*. Edited by Xing Ruan and Paul Hogben. London: Routledge, 2007, pp. 12–21.
26. James Joyce. *Ulysses*. Harmondsworth: Penguin, 1984, p. 11.
27. Peter Pearson. *Dún Laoghaire. Kingstown*. Dublin: O'Brien Press, 1981.
28. John de Courcy Ireland. *History of Dún Laoghaire Harbour*. Dublin: Edmund Burke Publisher, 2001.
29. Keith Basso. "Speaking with Names: Language in Landscape among the Western Apache." *Cultural Anthropology* 3 (1988): 99–130.
30. Kamal Boullata. *Belonging and Globalisation: Critical Essays in Contemporary Art and Culture*. London: Saqi Press, 2008.
31. Tom Vanderbilt. "Openings: Emily Jacir." *Artforum* 42 (2004): 140–41.
32. George E. Marcus. "Ethnography in/of the World System: The Emergence of Multi-Sited Ethnography." *Annual Review of Anthropology* 24 (1995): 95–117.
33. Doreen Massey. "Moving Spaces". In *Sensi/Able Spaces. Space, Art and the Environment.* Paper presented at the SPARTEN Conference, Reykjavik, Iceland, 1–2 June 2006. Edited by Edward H. Huijbens and Ólafur Páll Jónson. Newcastle: Cambridge Scholars Publishing, 2007, pp. 1–5.
34. See Iain Biggs. "Anthony Lyons' Quantock Dreaming: Developing a 'Deep Mappping' Practice." In *Weatherproof/Quantock Dreaming*. Bristol: Wild Conversations Press, 2012, p. 5. Available online: http://www.quantockdreaming.net/ (accessed on 31 May 2015).
35. For Hamacher's work in English, see Werner Hamacher. *Minima Philologica*. Translated by Catharine Diehl and Jason Groves. Bronx: Fordham University Press, 2015.

The Deep Mapping of Pennine Street: A Cartographic Fiction

Claire Reddleman

Abstract: Pennine Street is a cartographic art experiment, twinning High Street 2012 in London with the Pennine Way, a long-distance footpath running between the Peak District and the Scottish Borders. Pennine Street was initially prompted by the London 2012 Olympic spectacle; more specifically, by the militarization of the Games through the proposed deployment of surface-to-air missiles at sites in London. The project initially took the form of three organized walks along the route of High Street 2012, from Aldgate to Stratford. Readings were made while walking on each occasion, and both photographic and textual collages emerged out of the initial walks. The project engages the idea of trespass as a political action, as both potent and futile, and traces the development of modes of photographic and textual "trespass", or transgression. Textual collage is employed to investigate the possibility of articulating Pennine Street as a "space-between" the empirical and the imagined.

Reprinted from *Humanities*. Cite as: Reddleman, C. The Deep Mapping of Pennine Street: A Cartographic Fiction. *Humanities* **2015**, *4*, 760–774.

1. Introduction

Pennine Street is a cartographic art experiment, twinning High Street 2012 in London with the Pennine Way, a long-distance footpath running between the Peak District and the Scottish Borders. In this article, I give an account of this as-yet open-ended critical art project that considers it as a form of "deep mapping":

> a deep mapping occupied with, and attempting to give an account of, both a particular geographically located "place" and, simultaneously, a second space-between that both mediates and contests the exclusionary position of the authorizing official networks of culture and knowledge [1].

Pennine Street may be understood as constituting such a "space-between", that is able to mediate and to contest some of the dominant, "official" narratives that are currently functioning in the places of "east London" and the "Pennine Way". Where the concept and practice of deep mapping is still emerging and finding definition as a way of exploring place, I offer Pennine Street as a project that may contribute to our understanding of deep mapping, and that may deepen its attention to imaginative and fictional aspects of place.

In this article, I describe how the project of Pennine Street began and how it developed, attending to the personal experience of creative production as well as to the theoretical and political concerns that prompted and continue to animate the work. As I discuss further below, the invention and exploration of Pennine Street was initially prompted by the advent of the Olympics in east London. However, it is also more broadly concerned to explore, and problematize, the categorization of places through their (ontological and political) incorporation into new named entities, and the experience of perceiving and encountering such abstractions. Beginning from a juxtaposition of the two routes as contrasts to one another—rural/urban, established/new, traditional/postmodern—the project attempts to complicate and disrupt such binary formations, to explore how these formations inform and co-constitute one another, and how a theoretically-inspired approach to creative action might be able to play out as a form of political response to the spectacular politics of mega-events such as the Olympics, and the harms they cause. These are the motivations and the interests of the project, though of course I do not suggest that any of these issues are "answered" or resolved here.

The much larger political question of the effectiveness of protest is, for me, one of the most important unresolved questions of the Pennine Street project. With the conviction that conventional forms of direct action, demonstration and non-violent resistance would have no effect on the implementation and "success" of the Olympics, I made the also-ineffectual choice to develop a creative and theoretically-informed artistic project to attempt to address some of the underlying conceptual dynamics that make such phenomena possible; including considering how places can be defined and re-defined in this particular cultural moment; how the power to undertake processes of definition works through institutions and conventions such as naming; and how we might be able to re-consider what is accepted as "real" and "natural" through taking up an interest in fiction and imagination. These considerations have informed the development of Pennine Street as a resistant, imaginative response to the Olympic project, though they are, of course, not resolved in the project.

2. Critical Context

The project as a whole may be seen in the context of other approaches to re-thinking and re-experiencing place with an emphasis on radical political change. The Situationists, in particular, are a well-known antecedent in this area. The Situationist International was a group of artists, writers, architects and activists gathered around Guy Debord in the late 1950s, who wrote and collectively published a wide range of work loosely focussed on changing the experience and structure of everyday life in the city [2,3]. One of the most well known techniques the Situationists developed is the "dérive", a deliberate attempt to walk in the city while eschewing conventional way-finding, in order to reconfigure the received social constraints that are placed on ways of being and experimentally develop new ways of moving and new forms of consciousness. The initial walks of Pennine Street, in particular, owe some of their interest in walking as a form of political action to the example of the Situationists.

The latter, textual stage of Pennine Street draws on techniques used by members of Oulipo, the "workshop of potential literature", a collective of writers, mathematicians and scientists active from 1960. Working with writing constraints such as omitting particular letters of the alphabet and techniques for substituting words [4], the Oulipian concern to expand the formal possibilities for literature, and their commitment to experimentation and play, have influenced the technique of textual collage that I have developed in Pennine Street. The form of the Guidebook text, imitating the established guidebook genre, and working by means of the "constraint" of writing in quotation, is as important in the project as the content, the specific phrases, books and authors that are quoted. The techniques that I have used and developed in Pennine Street attempt to disrupt and undermine the Olympic re-casting of the A11 between Aldgate and Stratford as High Street 2012, to bring this route into a creative relationship with another significant British route, the Pennine Way, to explore both these routes and their formation, *and* the potential means through which we can understand the interaction between empirical and imaginative places as a political activity.

3. Walking "against" the Olympics

Pennine Street was initially prompted by the London 2012 Olympic spectacle; more specifically, by the militarization of the Games through the proposed deployment of surface-to-air missiles at sites in London. Its starting premise was to imaginatively, and performatively, "twin" two routes that I found to be personally significant and politically resonant at that moment in 2012: the so-called "High Street 2012" and the long-distance path known as the Pennine Way. High Street 2012 was an almost invisible phenomenon of the London 2012 Olympics. Re-branding the A11 between Aldgate and Stratford, in east London, High Street 2012 provided a rubric under which some refurbishment work was undertaken to the "public realm" along the route between the City of London and the Olympic site. By apparent contrast, the Pennine Way was established in 1965 as the first National Trail in Britain, allowing public access to a continuous stretch of footpaths, offering members of the public "a once in a lifetime experience" [5]. Running between Edale in Derbyshire and Kirk Yetholm in the Scottish Borders, the Pennine Way is 268 miles long and includes the most remote parts of the country to be designated as national trails.

The idea to "twin" these two very different routes emerged as my confused and wounded response, as researcher, artist and local resident, to the militarization of the London Olympics. A planned holiday to walk the Pennine Way became an opportunity to remove myself from immediate proximity to the Olympic spectacle while it took place, and to form an artistic response to both the political project of the Olympics and my own experience of it as a violating imposition.

Taking up the idea of twinned routes, the project initially took the form of three organized walks along the route of High Street 2012, from Aldgate to Stratford. Open to the public, the walks, named "Pennine Street Trespasses", were taken by small groups, who were invited to bring and read texts of any kind that they felt had a connection to the concerns of the project. (This participatory element of the project proved to be less successful than anticipated, and came to function as a prompt for my further reading and research rather than as a methodology that played a central role in the Pennine Street project.) The walks were explicitly framed as an act of both homage and parody of the Kinder Trespass of 1932, in which ramblers organized a Mass Trespass at Kinder Scout in the Derbyshire Peak District. The first Pennine Street Trespass took place on 24th April 2012, the 80th anniversary of the Kinder Trespass. This event plays a central role in contemporary narratives of struggle and political contestation of restrictions to access to the land, and through Pennine Street I attempt to open out a space to examine and challenge my own attitude to the Trespass, as well as its enrolment in somewhat triumphalist contemporary discourses of struggle.

I particularly wanted to complicate my own received impression of a simplified narrative in relation to the Kinder Trespass; based on partial awarenesses gathered as a walker, through reading websites and guidebooks, and consuming media such as newspapers and radio programmes, I wanted to make my own (partial, inadequate) process of impression-formation an object of inquiry and criticism as much as the "external" objects of inquiry such as the routes and their histories. Therefore rather than presenting an account of current attitudes towards, and uses of, the Kinder Trespass that aspired to intellectual objectivity, I was concerned to question my own perspective, particularly as someone without experience of any of the struggles for access to the private land on which the Pennine Way was established privileged position of being able to "escape" the Olympics.

I saw in the Pennine Way, and its establishment as the first "national trail" in Britain, a narrative of a constructed route as metonymic of the nation (the National Trails website and the official guidebook, also published by Natural England, both refer to the Pennine Way as following "the backbone of England" ([5]; [6], p. 12). In twinning the two routes, then, I was partly interested to position High Street 2012 in relationship to the Pennine Way as a cynical or inadequate national metonym for the early twenty-first century; where the Pennine Way has inspired hyperbole it also offers rich and rewarding experience, forming a problematic, questioning binary with High Street 2012 as offering no comparable opportunities.

The three walks followed the route of High Street 2012 between Aldgate and Stratford; the first and third walks proceeded west to east, and the second walk, east to west. Readings were made while walking on each occasion (to which I return, below). I took as the end of High Street 2012 the location of the final improvement project along the route; Stratford High Street DLR (Docklands Light Railway) Station (see Figure 1). This site is a short walk away from the main Stratford Station, which includes Overground and Underground trains and which saw extensive improvement works ahead of the Olympics. The distance between the end of High Street 2012 and the main station providing access to the Olympic site became amplified in the imaginary of Pennine Street, as a sign of the feeling of bureaucratic ineptitude, of plans being slightly off and unfinished, delayed. I connected this impression with my sense of the futility of physical, or legal, trespass; my feeling of powerlessness in the face of the Olympic spectacle. In this I also saw a certain problematic "sublimity" that I found myself attributing to the Olympics as a whole. In response to these considerations, I read the well-known "tilting at windmills" passage from *Don Quixote* by Miguel de Cervantes at the Stratford High Street DLR Station, and photographed the new lighting scheme:

Figure 1. Stratford High Street DLR Station, 2012.

A series of subsequent walks enabled me to photograph more of the route, and to extend the idea of trespass in the project through stickering. Stickers were placed along the route and its environs, offering waymarkers for the (non-existent) walker to follow (see Figures 2 and 3). On a personal level, I was surprised at how self-conscious I felt at making this very minor transgression, of placing stickers in public places, on signage and street furniture. I was not carrying out an act of trespass in the conventional and politicized sense of accessing land that is private or otherwise forbidden; rather, I considered that this attempt was to "trespass" in plain sight, so to speak, by unofficially and ineffectually declaring public spaces to be part of the new, non-official and fictional space of Pennine Street.

Figure 2. The Pennine Street waymarker sticker.

Figure 3. One of the waymarker stickers in position.

4. Trespass: Homage and Parody

In seeking to make a response of some kind to the imposition of the militarized Olympic spectacle, I took up the idea of "trespass" from two areas: the discourse of contemporary activism at large, in which trespass is often used as a form of resistance, and the particular event of historical trespass associated with the establishment of the Pennine Way, the "mass trespass" carried out at Kinder Scout in 1932. I was interested in this event and its subsequent uses, in terms of attempting to critique what I saw as its problematic contemporary deployment as a "success story" of popular political action, while at the same time admiring its effectiveness. The "Kinder Trespass" functions as both a justification and an inspiration for the "access movement" that continues today. This framing of the Kinder Trespass is exemplified by the Ramblers' online presentation of the 80th anniversary in 2012:

> On 24 April 1932, groups of ramblers left Manchester and Sheffield for an organized trespass onto Kinder Scout, a moorland plateau in what is now the Peak District. There they clashed with gamekeepers sent by local landowners to keep people off their land. The clashes were violent and several of the ramblers were arrested and imprisoned, but over the following days and weeks much larger trespasses were held and public opinion started to sway in the trespassers' favour [7].

Today, it is possible to trace the Kinder Scout trespass as the start of an access movement that saw the establishment of National Parks, long distance footpaths including National Trails and finally, the Countryside and Rights of Way Act 2000 which granted unrestricted access to 10,000 square kilometres of countryside in England and Wales. The Ramblers, as a national charity concerned with the promotion of both access to the land, and walking *per se*, also cites the Kinder Trespass as a central moment in its account of its own history.

In developing Pennine Street, I have wanted both to rail against the neatness and over-simplification of the deployment of the Kinder Trespass as an anchoring event in the narrative of "the access movement", but also to venerate it myself; to consider my own sense of gratitude for the fact of access that the Trespassers' activism and effort has generated and that is extremely important to me; and also to confront the tensions involved in performing a bourgeois subject position in relation to a movement that has been positioned as a working class victory against the figure of the bourgeois landowner, or more particularly, the departicularized figure of the "gamekeeper". Standing in for any actual landowner in accounts of the Trespass, the gamekeeper provides a coherent antagonist to the trespasser. Here again a paring-down, a critical simplification, of the presence and role of this figure takes place; for example, injuries to gamekeepers are cited in more in-depth accounts and left unmentioned in briefer and more widely circulated accounts, while the fact of imprisonment for the trespassers survives the simplification process [8].

I read in this narrativizing process a deep ambivalence toward the question of violence and violation in the Trespass as a political action. Physical violence (the gamekeeper's broken leg) is effaced in favour of symbolic violence—enacted at the level of the law rather than the level of the person. In Pennine Street, through positioning the initial walks as "trespasses" held in both homage and parody to the Kinder Trespassers, I attempted to perform some of my own sense of conflictedness about the importance of symbolic action as well as its non-effectiveness at the immediate political level. That is, I was very aware that any action of trespass or physical resistance to the Olympics that I could make would have no effect on their going ahead, but still felt that attempting to intervene at the level of the symbolic was important and possible. It was this sense of conflictedness itself that I attempted to evoke and explore through the idea of the walks as both homage and parody.

The visual register of Pennine Street also attempts to explore the notion of "trespass" through collage (see Figure 4)—"trespassing" the integral image in favour of images that disrupt and violate one another—appropriating imagery, and stickering (see Figures 2 and 3). The walks generated a lot of photographic imagery, which I experimented with using collage as a visual analogue of the twinning, combining and interspersing of places proposed by the project as a whole. These visual experiments enabled the incorporation of photographs from the Pennine Way, as well as the appropriation of parts of maps, particularly Google Maps (see Figure 5).

Figure 4. Early collage "sketch" incorporating photographs of the Kinder Plateau, Derbyshire and Mile End, London.

Figure 5. Early collage "sketch" incorporating satellite map image of Balderhead Reservoir, County Durham, and photograph of Mile End, London.

5. Declaring Twinned Places: Haworth Idea Store and Hadrian's Interchange

The experiments in collage gave rise to a more particular twinning of specific locations along the two routes, attending, though not rigidly, to relative position in relation to the route as a whole (see Figure 6). For example, the beginning of the route became "Aldgatedale", incorporating toponymic reference to both Aldgate in London and Edale in Derbyshire; the end point became "Kirk Stratford High St DLR Station", incorporating, in the same way, reference to Stratford High St DLR Station in London and Kirk Yetholm in the Scottish Borders. The need to map the routes more closely onto each other arose through making this more explicit correlation between the two routes.

Figure 6. Early layout bringing the two routes into a parallel orientation, and abandoning relative scale.

Particular twinned locations were "declared", including Haworth Idea Store (see Figure 7), Stepney-in-Ribblesdale, Hadrian's Interchange, Kielder-Stratford Forest and Hackney Heath. Haworth Idea Store was prompted by the emphasis placed on Haworth and Top Withins by the Pennine Way guidebook, *Pennine Way South: Edale to Bowes*, as key locations of interest to the Pennine Way walker, thus:

> Ahead and to the left is a broken spine of a wall leading to the ruins of Top Withins [...] Readers of Wuthering Heights may be disappointed by the first sight of this famous ruin, which has been "conserved", but the view from the lonely steading is still wonderful. After passing Top Withins the Way drops downhill along a well-worn literary trail ([6], p. 67).

Figure 7. Haworth Idea Store.

A further inset page is devoted to "The Brontës of Haworth" ([6], p. 69), and further describes Top Withins:

> Top Withins, the inspiration for Wuthering Heights, beckons the more committed. Its setting is splendid and it should stir the imagination, but one of the most attractive things about a ruin is that it is insubstantial. It is a pity to see the once-crumbling remains of Top Withins enshrined as a blockhouse ([6], p. 69).

The emphasis, as well as the guidebook's apparently straightforward treatment of a material ruin as the setting of a fiction, prompted the inclusion of "Haworth" as a place within Pennine Street. In the transposed mapping of the two routes, the "real" Whitechapel Idea Store in London occupies a geographical position within High Street 2012 that is roughly equivalent to Haworth's position within the Pennine Way; that is, about a quarter of the way along their respective whole routes. The two locations are twinned, and juxtaposed, as sites of official, or institutional, interest, the library and the heritage site positioned as sites for the improvement of their visitors, and both centrally concerned with "fiction". The resulting twinned site, Haworth Idea Store, was also photographed (see Figure 7).

A concern with criticism is in play throughout the project, and specifically here in the choice to draw a connection between Top Withins and Haworth, and the Whitechapel Idea Store, through the idea of fiction. An angry, sardonic tone, or attitude, may be read as functioning at all times in the project and its specific moments, performing an aggressive degree of criticism that is simultaneously proffered and disavowed. This may be read in the "Haworth Idea Store" in terms of a conflicted attack on the phenomenon of the Idea Store, with its avowed aims of "reaching the non-learning public" [9] while finding itself deeply entangled in contemporary politics of "place-making" ([10], p. 12) and the ongoing gentrification of Whitechapel. These developments are metonymized in the Whitechapel Idea Store, which comes to stand for the simultaneous improvement of services for the "non-learning public" and their social and economic marginalization. "Non-learning" in this context functions as a euphemism for working- and sub-working class, itself a deeply racialized category in Tower Hamlets. For me, as against the official view [11], the Idea Store is an ambivalent place of accessible learning dignified by designer architecture, on the one hand, and a place of emptiness and false promise on the other, offering a shockingly limited range of materials packaged for people whom the state (in the form of London Borough of Tower Hamlets) regards as having little value and scant potential for positive change. The concern with criticism in the project is partly, then, an anxiety about offering it at all toward a scheme widely regarded as successful; and partly with a self-reflexive awareness of the privilege of my own social and economic position in relation to those who are "targeted" ([11], p. 58) by such services as the Idea Store.

Hadrian's Interchange is the result of twinning the Bow Interchange in east London, a huge, multi-level, multi-lane gyratory, or roundabout, with Hadrian's Wall, once the fortified northernmost boundary of Roman Britain, now a world heritage site and major tourist destination. I discuss Hadrian's Interchange in more detail below.

The most recent photographic stage of Pennine Street is a series of photographic collages focussing on the new imaginary of "Hadrian's Interchange" (see Figures 8 and 9).

Figure 8. Collage depicting Hadrian's Interchange, composed of photographs of Hadrian's Wall and Bow Interchange.

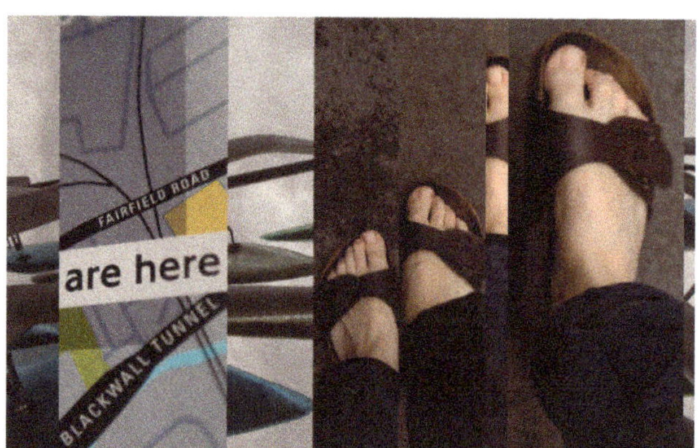

Figure 9. Collage combining photographs of walking at Bow Interchange, sculptural "trees" at Stratford, and navigational map signage.

6. "Deep Mapping" of Pennine Street through Textual Collage

The final major element of Pennine Street is a piece of fictional writing, parodying the form of the guidebook, and comprised almost entirely of quotations (see Figure 10). Initially, in wanting to form some sort of account of each Trespass walk, I drew on the texts that had been read out on the walk itself. Readings included the opening of *As I Walked Out One Midsummer Morning* by Laurie Lee, *The Country and the City* by Raymond Williams, *Gulliver's Travels* by Jonathan Swift, *The Return of the Native* by Thomas Hardy, and "Toba Tek Singh" from *Kingdom's End* by Saadat Hasan Manto. Readings were interspersed with extracts from the two-volume guidebook, *Pennine Way South: Edale to Bowes* and *Pennine Way North: Bowes to Kirk Yetholm*, both by Tony Hopkins.

> *An easy day rarely lives long in the memory, and the section of the* Pennine Street from the St Clement's Lahore Hospital to Hadrian's Interchange *has neither high nor low points to exhilarate or depress the walker.*[76]
>
> Hadrian's interchange *previously marked the edge of London for travellers leaving and coming to London. It now marks a transition between the traditional, urban and dense high streets of Tower Hamlets and the landscape of the Lower Lea Valley.*[77] *A vivid imagination is essential to see the Wall in its original glory, for nowhere is it even half complete. However, the setting helps to create a suitable atmosphere, and this should be an exciting moment [...] The deep fissures in these grey pavements are often green with moss and there are patches of wild chives – a rare plant said to have been introduced by the Romans. The Wall itself, after a fine turret standing up to 10 courses high, dwindles to a few fragments or the ghost of a mound.*[78]
>
> [76] Hopkins, Tony (2007) *Pennine Way South: Edale to Bowes*, Aurum Press and Natural England, p.82
> [77] http://www.highstreet2012.com/projects/area-initiatives/bow accessed 25.04.2012
> [78] Hopkins, Tony (2010) *Pennine Way North: Bowes to Kirk Yetholm*, Aurum Press and Natural England, p.92

Figure 10. Image of part of the Hadrian's Interchange section of the collaged text. The footnotes are intended to be highly visible and to function as part of the reading experience, both informing and interrupting the reading.

The guidebook became the central text for the project, due in part to my experience, when walking the Pennine Way using the guidebook, of Hopkins' narrative voice becoming something of a fictional presence throughout the walk. Providing a would-be objective and comprehensive description of what the walker will see, and need to negotiate (for example roads or areas of wet ground or poor signage), the narrative voice of the guidebook also habitually "deviates" into idiosyncratic reflections on points such as the conservation of Top Withins, noted above, or comments on the walker's "experience": "When you walk into Hawes early, after a comparatively easy walk, you will wonder what all the fuss was about." ([6], p. 104) The guidebook as both an object and a narrative voice becomes a central feature of the experience of walking. It is read aloud, and re-read, particularly when its description proves difficult to connect with the landscape itself; it is protected from rain, and held in the hand while walking. The author's idiosyncratic moments became moments to look forward to amid the succession of directions: "After crossing the lower meadows you begin to climb again" ([6], p. 120); "the route follows the towpath beneath a little bridge" ([6], p. 82); "just beyond this wood you bear right" ([6], p. 86).

As this text had taken up such a central role in my experience and memory of walking the Pennine Way, I used it to structure both the readings on the Trespasses and the written accounts I produced afterwards. Extracts from the guidebook describing, for example, one's view at a particular location, were interrupted with place-names taken from the map of east London. Extracts from the readings were also incorporated, drawing on the idea of collage used as a visual technique to "collage" the texts into new forms to create a fictional, imaginative "account" of walking Pennine Street—as though it were a "real" National Trail for which one could obtain a guidebook. The guidebook is structured by breaking down the whole route into manageable days of walking, for example, "4—Hebden Bridge to Ponden" and "5—Ponden to Thornton-in-Craven" [6]. I took up this format, sub-dividing Pennine Street into a recommended five-day walk between particular twinned, or combined, places:

Day One: Aldgatedale to the Haworth Idea Store
Day Two: Haworth Idea Store to St Clement's Lahore Hospital
Day Three: St Clement's Lahore Hospital to Hadrian's Interchange
Day Four: Hadrian's Interchange to Kielder-Stratford Forest
Day Five: Kielder-Stratford Forest to Kirk Stratford High St DLR Station

The length of High Street 2012 is around 4.5 miles, and the length of the Pennine Way 268 miles, and this mode of textual division holds together the scales of both routes without collapsing one into the other. The proposition that it could take five days to walk the length of High Street 2012 is, of course, absurd, potentially comic, as is the proposition that it would be possible to walk the Pennine Way in five days. A new sense of time and distance is suggested in the insistence on the scales and sizes of both routes.

The "twinned place" of Hadrian's Interchange offers an opportunity to consider in more detail how this place is evoked in Pennine Street through textual collage. The Interchange comes in the section "Day Three: St Clement's Lahore Hospital to Hadrian's Interchange" (see Figure 10).

Here, the first line comes from the guidebook by Tony Hopkins, *Pennine Way South: Edale to Bowes*, then from the "High Street 2012" website, then the guidebook again, with altered place names added by me. The idea of Hadrian's Interchange is of interest for me partly because of the sense, picked up in this passage, of a boundary or a point of transition, to the east of which is beyond what I think of as "my" east London. One crosses the River Lea, which is quite dwarfed by the structure of the roundabout and flyover. Thinking of this bleak urban space in conjunction with Hadrian's Wall gives an opportunity to consider the constitutive binaries of the military boundary, which was the Wall's original purpose, against the spatial boundary of the road and the river. Hadrian's Wall is now exclusively and relentlessly a site of leisure, and cultural consumption, but still controlled and protected. The guidebook's overbearing instruction on the need for a vivid imagination now comes to apply to both spaces; it retains its reference to Hadrian's Wall, and, placed in this new narrative context, it refers also to the incompleteness of the project of High Street 2012, which was originally part of the package of benefits that were supposed to accrue to the local area and the people who live there, but which were significantly diminished in the actual course of events.

There is both humour and performativity in the idea that this urban place could be proposed as somewhere that one might deliberately visit for the purpose of leisure, which as far as I can tell, no-one does. It is used as a shortcut by locals, but its layout does not provide for pedestrian access to the central space of the Interchange. The twinned place poses the further question, then, of what kinds of walking and access are understood as "leisure" and which kinds understood as "necessity" or everyday life. The need for a "vivid imagination", in this context, comes to refer also to the degree of imagination that may be required for those of us who are not compelled to live in this place to understand how it is done. The question of everyday life, and an imaginative engagement with how others live, was, I argue, absent from the project of High Street 2012.

The fictional Guidebook simultaneously offers an account and a guide to the imagined route, Pennine Street. The photographic collages act as illustrations to this imagined route, and are positioned at the relevant locations in the text, replicating the illustrative role of (non-collaged) photographic illustrations in the original guidebook. In this way, the capacity of photography to evoke the potential visual experience of the walker is disrupted through presenting altered images that "fail" to convey what may be seen when one is in place. The places that are combined in the images interrupt and supplement one another, resisting a coherent presentation of place. In the same spirit, the Guidebook's composition from quotations also stages this kind of interruption and deliberate lack of coherence. This mode of collaged writing came about in response to my desire to have the original voices and utterances of the "Trespass" texts appear directly, rather than being summarized or re-phrased, and my concurrent concern with learning and navigating the rules and conventions of academic writing as a PhD researcher.

The collaged text came to function as a way to experiment with taking up words written by others and my awareness of the violation entailed in the process of quotation, of making a selection from a coherent whole text, and re-presenting this selection in a new textual setting. I considered this approach in light of the initial concern with trespass, as a form of textual rather than physical trespass. I was also concerned to explore, in this specific context, the broader idea of how creating something new (whether a new named route, or a piece of writing) can entail dismantling and reconfiguring prior work by others. This method also apes, and subverts, the processes of forming and declaring the existence of the two routes engaged with in Pennine Street: both the more positive process of forming the Pennine Way through existing pathways, and the more superficial process of declaring High Street 2012 as a re-imagining of an existing main road.

While the project is currently a work in progress, the present version of the "guidebook text" may be accessed at the project's website [12]. I intend to publish the work as an artist's book, following the format of the guidebook with some of the photographic collages and hand drawn maps acting as guidebook illustrations. Further exploration of the ideas generated through the Pennine Street project could also take the form of replicating the twinning of two routes to create a third "space-between". In the same way that the declaration of the Pennine Way as a National Trail led on to the establishment of further National Trails, so the same process of twinning could be productively applied in other locations.

What I suggest is "mapped" in Pennine Street, then, in the expanded sense of deep mapping, is a layering of means of depiction simultaneously with two posited geographical locations. At

Hadrian's Interchange in particular, as with Pennine Street as a whole, textual collage is combined with photographic collage to evoke a "new" imaginative place, beyond the empirical; a "space-between". While Pennine Street continues to evolve, and therefore what is presented here is something of a report on a work in progress, I suggest that it indicates emerging possibilities for deep mapping to articulate and evoke imaginative aspects of place. In this way, approaches to place that we may consider under the rubric of deep mapping are able to push beyond the traditional cartographic concern with representation and engage with more political, and politicized, aspects of place and imagination.

Acknowledgments

Parts of this article were initially presented as a conference paper at the 2015 Nordic Geographers' Meeting, "Geographical Imagination: Interpretations of Nature, Art and Politics".

Conflicts of Interest

The author declares no conflict of interest.

References and Notes

1. Iain Biggs. "Deep Mapping as an Essaying of Place." Illustrated talk given at the Writing Seminar, the Bartlett School of Architecture, London, UK, 9 July 2010. Available online: http://www.iainbiggs.co.uk/text-deep-mapping-as-an-essaying-of-place/ (accessed on 1 November 2015).
2. The wide range of Situationist writings are usefully anthologized in *The Situationists and the City*. Edited by Tom McDonough. London: Verso, 2009.
3. See also *Situationist International Anthology*. Edited by Ken Knabb. Berkeley: Bureau of Public Secrets, 2006.
4. For a detailed anthology of oulipian innovations and techniques, see *The Oulipo Compendium*. Edited by Harry Mathews and Alastair Brotchie. London: Atlas, 1998.
5. National Trails. "Pennine Way." Available online: http://www.nationaltrail.co.uk/pennine-way (accessed on 1 November 2015).
6. Tony Hopkins. *Pennine Way South: Edale to Bowes*. London: Aurum Press, 2007.
7. Ramblers. "Kinder 80: Marking the Mass Trespass." Available online: http://www.ramblers.org.uk/kinder80 (accessed on 1 November 2015).
8. For an account of the Mass Trespass that is favourable to the Trespassers' cause, see Tom Stephenson, Ann Holt, and Mike Harding. *Forbidden Land: The Struggle for Access to Mountain and Moorland*. Manchester: Manchester University Press, 1989.

9. Heather Wills. "An innovative approach to reaching the non-learning public: The new Idea Stores in London." *The New Review of Library and Lifelong Learning* 4 (2003): 107–120. This article is authored by the Idea Store Programme Director and provides an institutional perspective on the programme. The Idea Store programme has been implemented in Tower Hamlets, the London borough which includes Whitechapel and much of the route of Pennine Street, ostensibly to rejuvenate the borough's provision of public library services.
10. Dorte Skot-Hansen, Casper Hvenegaard Rasmussen, and Henrik Jochumsen. "The role of public libraries in culture-led urban regeneration." *New Library World* 114 (2012): 7–19.
11. London Borough of Tower Hamlets. "Idea Store Strategy 2009." Available online: https://www.ideastore.co.uk/assets/documents/IdeaStoreStrategyAppx1CAB290709(1).pdf (accessed on 1 November 2015).
12. Claire Reddleman. "Pennine Street: A Cartographic Fiction in East London." Available online: http://www.penninestreet2012.wordpress.com (accessed on 1 November 2015).

Anticipating Deep Mapping: Tracing the Spatial Practice of Tim Robinson

Jos Smith

Abstract: There has been little academic research published on the work of Tim Robinson despite an illustrious career, first as an artist of the London avant-garde, then as a map-maker in the west of Ireland, and finally as an author of place. In part, this dearth is due to the difficulty of approaching these three diverse strands collectively. However, recent developments in the field of deep mapping encourage us to look back at the continuity of Robinson's achievements in full and offer a suitable framework for doing so. Socially engaged with living communities and a depth of historical knowledge about place, but at the same time keen to contribute artistically to the ongoing contemporary culture of place, the parameters of deep mapping are broad enough to encompass the range of Robinson's whole practice and suggest unique ways to illuminate his very unusual career. But Robinson's achievements also encourage a reflection on the historical context of deep mapping itself, as well as on the nature of its spatial practice (especially where space comes to connote a medium to be worked rather than an area/volume). With this in mind the following article both explores Robinson's work through deep mapping and deep mapping through the work of this unusual artist.

Reprinted from *Humanities*. Cite as: Smith, J. Anticipating Deep Mapping: Tracing the Spatial Practice of Tim Robinson. *Humanities* **2015**, *4*, 283–303.

1. Introduction

> "Imagine that in a few hundred years' time humanity has put aside all its misguided supernatural beliefs and turned its religious instincts to the Earth, the true author of our being. Then a rite will be called for to celebrate this thoroughly realist and romantic-materialist cult of the Earth. This rite will be the Visiting of Places, to contemplate them in all their particularity." ([1], p. 51).

This description of a "rite" associated with the secular contemplation of place presents a very telling knot in the author Tim Robinson's thought. He is, of course, an atheist who dismisses religion's "misguided supernatural beliefs" here. Such beliefs are, for him, too transcendental, too dependent on any other world than this. However, there is something lingering in the religious outlook that holds his fascination. Though religion itself is not for him, there is nothing wrong with our "religious instincts" which might simply be re-deployed into this "realist and romantic-materialist cult of the Earth". The subtitle of his first book is *Pilgrimage*, though it is a pilgrimage "with eyes raised to this world rather than lowered in prayer" ([2], p. 25). There is a pattern here. While he rejects the supernatural, there lingers a ritualised exploration of place suggesting veneration for its hidden dimensions, for its more-than-immediately-apparent meanings. As such, place still presents a metaphysical question for Robinson, but a question about immanent rather than transcendent metaphysics. The key to understanding Robinson's distinctive, secular metaphysics, however, is in his fascination with space, not place: "I prefer this body of work to be read in light of 'Space' [...] somatic space, perceptual space, existential space and so on," he describes, "ultimately there is no space but Space, 'nor am I out of it', to quote Marlow's Satan, for it is, among everything else, the interlocking of all our mental and physical trajectories, good or ill, through all the subspaces of experience up to the cosmic" ([1], p. vi).

Place and landscape are the themes with which his writing is more readily associated, grounded in that "particularity" he mentions above. However, place and landscape might simply be the most concrete ways to apprehend the more labyrinthine and metaphysical exploration of space itself, one that extends deep into perception, subjective experience, memory, community, history, knowledge, and even a variety of disciplinary expertise. One way of thinking about deep mapping is as a form of *working* the social and cultural medium of place—carefully and deliberately manipulating the relationships between these various and "interlocking" depths and fields that our social life opens up—and in this sense the medium of place, for Robinson, is "Space". There are a great many definitions of space and place to draw on today from phenomenological to Marxist geographers but this article draws its definitions from Robinson himself in reference to his own distinctive spatial thinking. Contrary, for example, to a philosopher, such as Edward S. Casey, for Robinson, it is "Space" that is primal in our universe, not place. Whereas, for Casey, space and time are concepts abstracted out of the immediate place-world, for Robinson, place is preceded by "Space"; it contains *and* is contained by "Space" [3]. I am not arguing necessarily for the truth of this view, but rather setting out, at the beginning, the distinctive spatial premise on which Robinson's thought it based and from which his wider practice emerges.

Robinson is not an author alone, and before he was an author he was a maker of maps. But before he was a maker of maps, he was also an artist practicing under the name of Timothy Drever, preoccupied with abstract and geometrical forms and their relationship with the spaces of exhibition and the art-going public, with topographical and with social space. In more ways than one then, space precedes place in his work. When we understand his career as more than writing alone, as a practice that has art at one end, map-making in the middle, and writing at the other, we see that space itself is the *medium* with which he has been working all this time.

Robinson was born in Yorkshire and studied mathematics at Cambridge before leaving England to travel and explore a career in visual art, living for periods in Istanbul, Vienna and London. However, at the very point when his career as the artist Timothy Drever was gathering momentum in London, he left, turning his back both on the city as a cultural centre and on visual art as the medium in which he would work. In 1972 he and his wife moved to the Aran Islands off the west coast of Ireland in the mouth of Galway Bay where they lived for over ten years before moving once again to the village of Roundstone on the coast of Connemara where he still lives today. Since this dramatic move west he has created several editions of new maps of the Aran Islands, the Burren, and Connemara (what he calls his "ABC of earth-wonders") ([1], p. vi). The Aran maps in particular were the first to have been made since the Ordnance Survey had attempted to do so as a part of the British government's colonial administration in the mid-nineteenth century. When he began to produce these maps it was out of an innocent curiosity as to whether it might be possible "to make amends", correcting the misplacement of topographical features and the mistranslation and Anglicisation of Gaelic place names that had taken place under the British and in which he had detected a "carelessness that reveal[ed] contempt" ([1], p. 3). The research undertaken for the maps soon led to what he called the "world-hungry art of words" and he is now the author of a two volume study of the Aran Islands a three volume study of Connemara, two editions of miscellaneous essays mostly exploring the same places, and a volume of short fiction and experimental writings. These works, as John Elder has suggested, have earned Robinson "a permanent place on the shelf that holds the scientifically informed, speculative and at the same time highly personal narratives of such earlier masters as Gilbert White and Henry David Thoreau" ([4], p. 1).

The relationship between his map-making and prose is well known and something that he has reflected on frequently in the books. However, there has been little reflection on the relationship between all *three* aspects of his career—the maps, the writing, and the earlier visual art[1]. This is despite his having exhibited at some quite prestigious galleries in the 1960s and 1970s[2], his selection for exhibition at a John Moore's Biennial in Liverpool by a panel of judges featuring Clement Greenberg in 1965 ([5], p. 44), and despite there even being a print of his still to be found in the Tate Gallery archives in London today [6]. By looking a little more closely at the relationship between this neglected alter ego, Timothy Drever, and the more familiar Tim Robinson, a comprehensive picture appears of the developing practice of a quite singular, inquiring mind, one capable of very striking leaps of faith in pursuit of his elusive subject matter. "Deep mapping" as an emerging form of artistic and academic practice offers perhaps the most appropriate critical framework through which to consider Robinson's work like this *in full* without segregating any one aspect of the work such as his writing, as other studies have (very illuminatingly) done [7,8]. Deep mapping may also be the only appropriate description of precisely what he has been doing in the west of Ireland since 1972.

The early work of Tim Robinson that this article discusses took place largely before the (practice-led) theorisation of "deep mapping" by figures in Europe such as Ian Biggs [9,10], and Mike Pearson, Clifford McLucas, and Michael Shanks [11]. It even precedes the moment when the idea of it was called forth from the work of Wallace Stegner by William Least Heat-Moon [12] in the American tradition. Nonetheless, the way that deep mapping, in the UK especially, has formulated itself *in practice* does suggest a reflection on the curious historical relationship between it and Robinson's work. Not least of all because, for Ian Biggs in particular, Robinson is singled out as a figure who "anticipates" deep mapping, part of a thread that Biggs traces back through John Cowper Powys to Thoreau ([10], p. 11). Reading Robinson's work holistically as "deep mapping" rather than just as writing also encourages us to draw attention to aspects of the work that have involved a wider, committed social engagement, the germ of which emerges from certain tensions that were there in his early artistic practice. Deep mapping is at heart a form of place-making, or place-transformation. It recognises that the identity associated with place is not a matter of essence, stability, and boundedness but of work, life, and creative energy. It explores new dialogues between the

---

1 The exception to this an autobiographical essay in *My Time in Space* and the slim volume issued from Coracle Press in 2012 called *The View from the Horizon*, both of which I will be drawing on here.

2 The Walker Art Gallery, Liverpool (1965), Signals Gallery, London (1966), Lissom Gallery (1968), Kenwood House (1969).

variety of perspectives with which a place is invested, past and present, though with an emphasis on "constructive reconciliations in the present" ([9], p. 5).

The terminology associated with the practice of deep mapping—and it is crucially to be understood *as practice*—can be read along a continuum of verbs that enact an engaged cultural work associated with this transformation. Drawing on Biggs again, deep mapping "intervenes", "challenges", "destabilizes", "mediates", and "reconfigures" "existing territories and presuppositions" ([9], p. 5). It offers a form of resistance to prevailing conventions of place representation and a recovery of the rich but underappreciated cultures going overlooked. As such it has to it a fundamental inclusiveness of attitude and often a quite radical "heterogeneity" of outcome ([11], p. 166). This is often a matter of nurturing the less visible "small heritages" (to borrow an eloquently plural term from David Harvey)—that can be anything as small as a field name, a superstition, or pneumonic—by a range of means ([13], p. 20). Tim Robinson's body of maps and written work, understood within the wider context of his whole artistic practice, can be read along a similar continuum insofar as they have endeavoured to engage with an intangible and oral topographical culture damaged by a colonial legacy and often found to be precariously balanced on the edge of memory.

Biggs has described deep mapping not just as interdisciplinary but as "challenging the very distinctions between academic and artistic outcomes" ([9], p. 5). In doing so, academic research and artistic practice often work collaboratively to produce what he calls "a potent catalyst for social change" ([9], p. 5). As will become apparent, Robinson's artwork, maps and the accompanying books represent a sustained form of cultural heritage and conservation work that operates along very similar lines. His achievements have now been celebrated by the people of the places themselves, by a highly commended citation in the British Cartographical Design Awards in 1992, and by a European Conservation Award, recognising the work of his and his wife Máiréad's company Folding Landscapes, in 1987 [14]. These are significant forms of recognition for a socially engaged practice of place-based heritage work that goes beyond the literary achievements for which he is best known.

In what follows, I explore Robinson's preoccupation with space through the three key aspects of his career. In the first part, his early experiments with the geometrical spaces of autonomous abstraction in a particular strain of modernism in the visual arts are shown to expand in conflicted ways out into the public and social sphere at a crucial time of political awakening in London in the 1970s. This conflicted expansion is then shown to be the guiding influence in his navigation through this very unusual exploration of map-making in Ireland which begins to challenge the conventional sense of cartographic space and explore more inclusive and non-standardised forms. In the third part, this same line of developing spatial thought and practice is traced from cartographic to linguistic forms of space as language is found to be able to do what neither the art nor the map could. The article concludes with a reading of a particular prose aesthetic of "consilience" in the written work which, it is argued, can be understood as the culmination of Robinson's spatial practice and thought. Finally, I argue that his work not only anticipates deep mapping but also exceeds it and extends it in a distinctive, literary direction. While a reflection on this career-long development of a line of spatial thought and practice will offer a uniquely full reading of Tim Robinson's work, at the same time, it aims to reveal something about the wider kinds of twentieth century cultural history out of which "deep mapping" might be understood to emerge, as well as looking toward the future in offering a reflection on the nature of its wider spatial practice where space itself becomes a social and artistic medium.

2. The Art: "A Bridge into the Real World"

In 1996, Robinson was asked to take part in an exhibition at the Irish Museum of Modern Art in Dublin. It had been nearly twenty-five years since he had left the London art world and turned to map-making and the literary essay. Nonetheless, the work he chose to exhibit brought together his earlier visual art with his mapping and writing in an interesting way that demonstrated a certain surprising coherence of thought. In the middle of the room, scattered on the floor like large pick-a-sticks were what seemed to be surveyors' rods, some with equidistant black and white stripes, some just white with a single inch painted grey at different points on the rods, and above them, suspended by a splayed rainbow of thread, was one more yard-long white rod. The lines on the black and white rods were not, however, all equal, suggesting a certain divergence from the standard that they brought to mind. On the walls around them were two of his intricate, hand-drawn maps of the Aran Islands and of Connemara, and between them were some twelve extracts from his books *Stones of Aran: Pilgrimage* and *Setting Foot of the Shores of Connemara and Other Writings* [15]. After visiting the exhibition, a friend described the surveyor's rods on the floor as "measure become organic" (quoted in [15], p. 11). It is an interesting phrase in which there is a sense that the measure has somehow lapsed or that it has been overcome from the inside. The phrase has an echo of "gone native" to it, since what use is measure if it is not answerable to a universal standard? There is something absurd and paradoxical about these surveyors' rods, each with its own measure and none of them bound by the same proportions. The white rods with a single inch painted grey at different points were called "Inchworm", a name for the caterpillar form of the geometer moth, so called because, according to Robinson, its movement in small loops seems to "measure the Earth" ([15], p. 57). Again, there is something absurd about the idea of an animal that might measure to no purpose other than travel. The measurement is not recorded or abstracted but simply performed. Life as lived is the only measure of which these rods speak. They *are* a standard rather than appealing to one. There is something very strangely prescient in this installation, the rods of which, were created originally before Robinson left London in 1972 and had been in storage all the while. They seem to have within them the kernel that would grow into his remapping of the Aran Islands, the Burren and Connemara, refusing the standards of the nineteenth century Ordnance Survey and asserting a form of spatial autonomy.

The level of abstraction and the subtle but philosophical commentary on space that we see in the rods here was characteristic of Robinson/Drever's London work of the time, though of course without the accompanying maps and writing. In the 1960s, modernism had returned to the London art world reconstituted by American intellectuals like Clement Greenberg who, since 1939, had been defending a purist abstraction and the avant-garde in a way that we might imagine could appeal to Robinson's background in mathematics. For Greenberg, abstraction narrowed and raised art "to the expression of an absolute" in which "subject matter or content" had become "something to be avoided like a plague" ([16], p. 531). This led, he suggested, to "free and autonomous" work, pure painting or sculpture, "valid solely on its own terms" ([16], p. 531). As mentioned, Greenberg had been among the judges who selected Robinson/Drever to exhibit in the John Moore's Biennial in Liverpool in 1965 and other exhibited works of his from around the same time also show a fascination with geometry and mathematical proportions. For example, the print that remains in the archive of the Tate Gallery in London is one, the form, composition and proportions of which were produced by a strict adherence to certain geometrical principles and rules (see Figure 1). As he describes in an exhibition catalogue from the time, "aesthetic choices were progressively replaced or limited (and so made more crucial) by geometrical demands" ([17], p. 15).

Figure 1. Timothy Drever. Untitled. 1969. Tate Gallery.

However, there was also an emergent pull away from the "autonomy" of abstraction at this time and it is this subsequent tension between the two that would propel him out of London in 1972. For an exhibition in the summer of 1969, he and the artist Peter Joseph published an essay in *Studio International* called "Outside the Gallery System". In it they voiced their dissent at an art world bound up with commodity fetishism, suggesting that this "increasingly isolates the artist from the public", channeling work "at best into a museum, at worst into an investor's cellar", leaving the artists themselves to a "comfortable enervation" ([18], p. 255). Robinson/Drever and Joseph set about challenging this by holding their exhibition outdoors in the grounds of Kenwood House. Not only this but the art they exhibited relied on the interaction and participation of visitors to be fully realised. "Consideration of the environment is essential", they declared, "the scale and dynamics of the work must relate to the area in which it is shown. Thus, it seems natural that 'environmental art' should be not just the latest fad of the art-world, but a bridge into the real world" ([18], p. 255). Robinson's own exhibition piece was a series of large, flat, coloured shapes produced, again, according to geometrical rules, but here they were set down on the lawn and he invited the public to move, reorder, and experiment with new compositions to bring the work alive (see Figure 2). This echoed a previous interactive installation that he had exhibited indoors at Kenwood House the same year called "Moonfield" where visitors were invited to enter a darkened room with a black floor only to find their feet knocking black wooden shapes on the floor which they were encouraged to turn over revealing a white underside. As people made their way through the room new patterns of black and white emerged dependent on their physical interaction. "Moonfield is thus a new surface to be explored, and this exploration creates its topography", the exhibition catalogue claimed ([17], p. 5). This topography that was created through the interaction between artist, artwork, and public was described as a "real space both in philosophic and social terms" ([17], p. 17).

Figure 2. Timothy Drever. Four-Colour Theorum. Kenwood, 1969.

Both of these works demonstrate the fascinating paradox of an artist working in forms of autonomous abstraction while at the same time expressing a yearning for the social engagement and interaction that abstraction had spurned. In the case of the outdoor exhibition at Kenwood House, there was also clearly a disillusionment emerging with the conventions of the gallery-oriented metropolitan art world. Deep mapping itself can be seen to emerge from a similar uncomfortable feeling about the lines drawn between an "art world" and a "real world" though it has also grown to challenge the lines separating an "academic world" as well. On the one hand, this "environmental art" was not then what we might expect it to be today, "environment" referring simply to the nature of the public space of exhibition. It was simply about finding, quite concretely, new environments for art and encouraging real human interaction through which the visitor takes part in the process of creation. However, the decision to search for a new environment does speak of a frustration with the prevailing spatial discourse of exhibitions and recognition of the need to question its hegemonic hold on the relationship between artist and the public. This seems to parallel a growing realisation at the time of modernism's own waning political antagonism. Alan Sinfield reminds us how easily Greenberg's defence of the autonomous freedom of the abstract expressionist was co-opted into an ideology threading through a number of C.I.A.-funded European exhibitions that served as propaganda in the Cold War ([19], pp. 210–14). By bridging the art world and the real world, or rather by making that bridge out of the art world and "into the real world", connecting the gallery with wider, more democratically public, environments, and with questioning the social and economic implications of setting a work in a metropolitan gallery, Robinson himself has noted that there may have even been the beginnings for him of an "environmental art" in the sense that it is more readily understood today ([20], p. 4). In his essay "Environments" in that same edition of *Studio*

International in 1969, the performance artist Stuart Brisley asks: "[t]o what extent does the artist maintain responsibility for the implications implicit within his artistic processes beyond production?" suggesting that the commodification of art ought *not* to be something of which the artist passively disapproves while continuing to feed ([21], p. 267). For Brisley, as perhaps for Robinson/Drever at that time, "environmental work specifies that the artist take a positive position in relation to his own behaviour as it affects other people within the social and physical context" ([21], p. 268).

It was during this same period that Clarrie Wallis recalls the first "walk-as-art" taken by the young Richard Long and Hamish Fulton, a conscious decision to turn their backs on the city as authoritarian centre of culture. She quotes from Long's diary of 1967:

> We announced we were going to walk (at a normal pace) [...], out of London until sunset. A few didn't start. We went along Oxford Street to the Edgware Road—the Old Roman road of Watling Street—which we followed in a more-or-less straight line north-west out of London. A few more people dropped out along the way, leaving about six of us at the end. We had no preconceived idea of where we would end up; in fact at sunset we found ourselves in a field, not lost, but also not knowing exactly where we were. The first place we came out to was Radlett, so we caught a train back from Radlett station. (quoted in [22], pp. 42–43).

Here, the walk-as-art appears an eccentric student experiment influenced no doubt by student marches of the earlier 1960s. However, there is something very carefully thought out about this, the spatialization of a train of thought or feeling that encourages us to read this walk as a performance, even a tenuously constituted sculpture. Wallis has gone so far as to describe this moment as representative of a "shift in consciousness [...] the end of Greenbergian modernism and the beginning of a new era. It coincided with a turning point away from technological optimism to preoccupations with ecology, conservation and a crisis of the 1970s as the British were uneasily forced to face their post-industrial and post-colonial future" ([22], p. 38). In many ways this walk, through the very area where Robinson was living at the time, can be seen to parallel his own departure from London for Aran along a similarly north-westerly axis just a few years later[3]. Both moves offer a performative rejection of the capital at a key time, a rejection of what Raymond Williams has critiqued in modernism as the "persistent intellectual hegemony of the metropolis" ([23], p. 38).

It was after seeing Robert Flaherty's film *Man of Aran* (1934) that Robinson and his wife decided to make the move. The film is a strange blend of documentary anthropology and dramatized narrative, but it was the film's sheer Romanticism that appealed to them at the time: windswept cliffs, a life lived between rough limestone and the Atlantic Ocean on the outer edge of Europe. There could have been nowhere further from the modernist metropolis. Nonetheless, though such Romanticism might have been the initial appeal, it should be noted that Aran became

---

3 Robinson has described himself at the same time taking "long abstracted country walks", oriented "by glimpses of the spires of Kilburn, Cricklewood and Neasden" all in fact districts collected around that same ancient trackway of Watling Street ([15], p. 55). Like Long and Fulton's navigation by an ancient trackway, Robinson's navigation by church spire speaks of that same search for an older orientation as a foil to the modern urban architecture around him.

for them not an escape *from* the world so much as an escape *into* a new world. Theirs was not a move of retreat from social and cultural politics and Robinson would be drawn to the complex political, international, and economic tensions that ran through the life and history of the islands just as J.M. Synge had before him. When Yeats offered that famous advice to Synge in the early twentieth century—"Give up Paris [...] Go to the Aran Islands. Live there as if you were one of the people themselves; express a life that has never found expression" (quoted in [24], p. xix)—Synge found a way of life and culture richly alive with the language, history and folklore that had already become so important to a burgeoning sense of Irish national identity. He found, in the cottage of Páidin and Máire MacDonncha, not rural isolation but what had come to be known as *Ollscoil n Gaeilge* (the University of Irish) due to the number of scholars that had stayed there for their research ([25], p. xvii). Synge also found it a very poor place, exploited by ever rising rents set by legislation decided upon in Westminster, a place in which people were being evicted from their homes ([25], p. 143). Though Robinson had not read Synge before he left, this sense of the Aran Islands as a centre of culture in their own right, connected by a long and complicated history to Britain, was something he would very soon come to discover.

In the first few years of living on the islands, Robinson attended an exhibition of Richard Long's in Amsterdam and found on the poster one of Long's sculptures photographed on Aran in 1975. Long had also spent a summer there but they do not seem to have met at that time [26]. He also describes himself and his wife catching sight of one of Long's sculptures from a plane window when they were returning after a trip away, "an instantly recognisable mark that told us who had visited the island in our absence" ([2], p. 44). In fact, he actually gets into a dispute when Long is "aghast" to find two of his stone-works marked on Robinson's map of the island (quoted in [27], p. 113). "[T]he essence of his works", Robinson concedes, "is what he brings home to the artworld: a photographic image in many cases" which serves as "the entrypoint to a concept, the idea of a journey" ([27], p. 115). Robinson reminds Long that, nonetheless, he does in fact leave something behind, something that he, as a maker of a very detailed map, could only find it hard to ignore. The tension is here once again between abstraction ("the entrypoint to a concept") and a more tangible real-world social interaction (the stone-work left behind). For Long, the pull back to the metropolitan centre from the periphery was still strong and where the work reached its final realisation, but for Robinson, who was by that time beginning to map the islands and recognise them as a centre in their own right, that "bridge to the real world" was coming to look final.

3. The Maps: "Making Amends"

On the south west coast of Aran, at the base of its two-hundred-and-fifty-feet limestone cliffs, there is a cave called An Poll Dubh, or "the black hole", in which a piper is said to have wandered, never to be seen again. The folklore of the islands has it, though, that inland, under the village of Creig an Chéirín, his music can still sometimes be heard. It is a story that occurs across England as well, though sometimes in the guise of a fiddler rather than a piper. Jennifer Westwood and Jacqueline Simpson tell of a group of people in the village of Anstey in Hertfordshire, for example, who were curious about how deep a local cave went so they sent a fiddler below ground, playing his fiddle, while they walked above ground listening as a way of plotting the depth of the cave in the known landscape. In a horrible twist, the music stops suddenly as the fiddler is taken by the Devil and never heard of again while his dog comes running from the cave with its hair burned off ([28], pp. 332–33). Similar stories are associated with Grantchester in Cambridgeshire, Binham Priory in Norfolk, Peninnis Head in the Isles of Scilly, and Richmond Castle in Yorkshire. The story becomes something quite different in Robinson's hands. Though it is not clear whether this is because of its particular inflection in the memory of the islanders or because of a certain license he takes himself (more than likely it is both), the sense of a community collaborating with the piper is stripped away and he is reported to be exploring the cave alone. The function of his music in the story becomes a little more mysterious too, as in Aran it shifts from a means of mapping the cave to merely a haunting vestige of a disappeared man. Robinson absorbs this story into a personal mythology in *Stones of Aran: Pilgrimage* and we get a curious glimpse of the lone, questing artist, literally immersed in place, but somehow searching for something more mysterious within it as well:

> Thus: the artist finds deep-lying passages, unsuspected correspondences, unrevealed concordances, leading from element to element of reality, and celebrates them in the darkness of the solipsism necessary to his undertaking, but at best it is a weak and intermittent music, confused by its own echoes and muffled by the chattering waters of the earth, that reaches the surface-dweller above; nor does the artist emerge; his way leads on and on, or about and about. ([2], p. 129).

It is an image in which we can certainly read feelings about his remote and isolated existence in those early years on the islands, perhaps even a certain alienation from the place itself, coupled of course with a fear of being literally swallowed by it. And yet, there is a paradox that emerges in this image as well. The "darkness of the solipsism necessary to his undertaking" is, in fact, not a darkness of solipsism at all. It is a darkness that belongs to both the island's geology and its folklore. Robinson's account of a supposed interior, subjective space is contradicted by its origin in the island's own cultural repertoire, the recognition of what was, in fact, a shared cultural space. Though the community who trace the piper's movement from above ground are stripped away from this story then, nonetheless, in the "weak and intermittent music" that does make its way to the "surface-dweller" we can see the beginnings of a tentative emergence of the artist to his social context. The withdrawn autonomy of Drever the modernist artist was beginning to unfold his fascination with abstract space into the manifest reality of the island's cultural landscape as it was encountered by Robinson the hoarder of place lore.

He had begun collecting the place names, stories, and histories like this after Máire Bn. Uí Chonghaile, the postmistress of Cill Mhuirbhigh, suggested that he make a map of the islands. He was surprised to find that no map had been made since the Ordnance Survey's nineteenth century project which had produced impractical, large scale maps, copies of which were difficult to get hold of. He did, nonetheless, acquire copies and began studying them. He learned Gaelic, still then the first language of the islands, and spent a lot of time visiting his neighbours, walking with farmers and fishermen, and talking at length about the topography and its associated history and culture over tea. He soon found that the maps were full of errors, not just of mistranslation and Anglicisation but also of misattribution. As he began to collect place names and understand the stories and histories associated with them, the rich depths of life and heritage that had been ignored or misunderstood, and that often existed only in the memory of a few ageing individuals, he began to understand that, for the officials planning the survey from Westminster, "rents and rates came before any other aspect of life" and for many of the soldiers conducting the survey on the ground, the "language of the peasant was nothing more than a subversive muttering behind the landlord's back" ([1], p. 3). We can imagine the appeal of correcting these mistakes, and of exploring the possibilities of reparation that might be involved with representing, for the first time, the actual geographical complexities he was discovering in getting to know his neighbours. He was, after all, an artist emerging from a movement that Stuart Brisley had described as in search of that "positive position in relation to his own behaviour as it affects other people within the social and physical context" ([21], p. 268).

Looking at early cartography up to the medieval period, Michel de Certeau reminds us that there was once a much closer relationship between the map and linguistic description than there is today. Early maps were often written on with accounts of tours, histories, or itineraries of pilgrimages ([29], p. 121). Their two dimensions offered up stories of the mapmaking process; but, he says, these stories were slowly shouldered out to make way for more purely visual-spatial description. The history of the map "colonizes space; it eliminates little by little the pictural figurations of the practices that produce it" in favour of the top down, precisely surveyed representation of static space dotted with symbols that we have today ([29], p. 121). This erasure of the stories of the landscape from the official representation is nowhere more felt than Ireland. Too small to have developed its own map-making tradition before the English came with theirs, Ireland did nonetheless have its own rich linguistic geography: *dinnseanchas*, the oral tradition of keeping the lore of the land. Charles Bowen describes *dinnseanchas* as "a science of geography [...] in which there is no clear distinction between the general principles of topography or direction-finding and the intimate knowledge of particular places" ([30], p. 115). He goes on: "Places would have been known to them as people were: by face, name and history [...] the name of every place was assumed to be an expression of its history" ([30], p. 115). Unlike the increasingly spatial mapping practices of the English, this method had a temporal and historical depth to it and existed, not on paper, but in memory and social interaction.

From the 1520s the English government began commissioning maps of Ireland. Begun as they were, just before such a trend of recording linguistic description began to die out in the 17th and 18th centuries, these first maps do in fact contain a few of these examples of the kinds of place lore Bowen refers to actually written onto them in the style of the medieval maps that de Certeau describes. J. H. Andrews tells us that on the Dartmouth Maps of 1598, for example, there is the description: "O'Donnell camped by this logh where his men did see 2 waterhorses of a huge bigness" ([31], p. 202). Or the following even stranger piece from the same map: "In this bog there is every whott [hot?] summer strange fighting of battles sometimes at foot sometimes wt horse, sometimes castles seen on a sudden, sometimes great store of cows driving & fighting for them" ([31], p. 202). This was, however, the exception to the rule and such curiosities should be read alongside derisive illustrations of "wild Irishmen peeping from behind rocks" and in the context of a brutal colonial rule ([31], p. 202). Additionally, as Andrews explains, what there was of this practice soon died out as main roads were introduced and maps of Ireland began to endeavour to be more objective for the purposes of administration, achieving a certain regimented abstraction. A concern for the Irish *dinnseanchas* would not be seen until briefly the Ordnance Survey set up its Topographical Department charged with the collection of heritage information in 1835. Even this, however, was brought to an end in 1842 on the basis that it was "stimulating national sentiment in a morbid, deplorable and tendentious manner" (quoted in [32], p. 287). In the English mapping of Ireland a certain living history was erased from the map before it ever really found its place on it.

For Robinson, the work that began to present itself was a matter of collecting and identifying the correct place names, representing them on his map in the correct place, and then subsequently recording the stories associated with them in supplementary written material. The map and the book together seemed the only logical way of making a record of a place so linguistically alive. Patrick Curry has called this "a kind of Edenic naming in reverse", a recovery of a world beneath the English language that had imposed itself ([33], p. 13). In an interview Robinson describes a typical example of the kind of work he was beginning to undertake:

> A very striking case was a place name that was recorded down at the south-eastern corner of the big island. It was something like "Illaunanaur". The surveyors had obviously thought that the first part of it was "oileán", island, when in fact it should have been the Irish "glean", glen. But apart from making it an island when it was a glen, the rest of the name "-anaur" meant absolutely nothing in English phonetics. But in the Irish the name means "the glen of tears"—it's exactly the biblical phrase "this vale of tears", "Gleann na nDeor". And the story I heard from the local people, was that, in the days leading up to the famine when there was a lot of emigration from the islands, those emigrating would get a fishing boat to take them over to Connemara and they'd walk 30 miles along the Connemara coast into Galway, where they'd wait for one of the famine ships heading for America. These ships used to sail out past the Aran Islands and very frequently had to wait in the shelter of the islands while a gale blew itself out. So they would be stationary just a few hundred yards off shore from this place, Gleann na nDeor, and people would come down to that little glen where they could wave to their loved ones but not talk to them. So the name had immense resonances and told you an immense amount about the personal griefs behind the statistics of the famine. ([20], p. 6).

It is not at all unusual that such a small name as Gleann na nDeor should contain such an elaborate and interesting story. Thinking of this coastal place as a "vale of tears", a phrase used in Christian theology to describe the world that must be endured before the soul can pass on to Heaven, suggests a poignant sense of hope about the life that might await relatives making their way to America. Yet unmapped, such names were slipping out of memory and there are numerous examples of intriguing names that Robinson is unable to find an explanation for. He describes this work as a kind of "rescue archaeology", gathering things in from the outer edge before they are lost forever ([1], p. 13).

As for the form of the map itself, he set about exploring something that would be importantly founded on the place itself. In a sense he was liberated by being able to tailor his map to so specific and small a location. The rules he worked by did not have to conform to so abstract or generalised a standard as those of the Ordnance Survey. For example, on its south-west side Aran is all cliff and on its north-east all beach, so the angle of vision that looks down on the island in his map is tilted slightly to the south west—what he calls a "seagull's-eye perspective"—thereby capturing the shapes of the sea-cliffs (see Figure 3) ([34], p. 38). This was important to a fishing culture that navigated by these shapes, and that had their own names for many of the headlands that differed from the inland names that farmers had for the same features. By making his such an isolated study, Robinson was enjoying a freedom of singularity, beholden to no distant, administrative standard, something that of course recalls the surveyors' rods he had created in London in 1972.

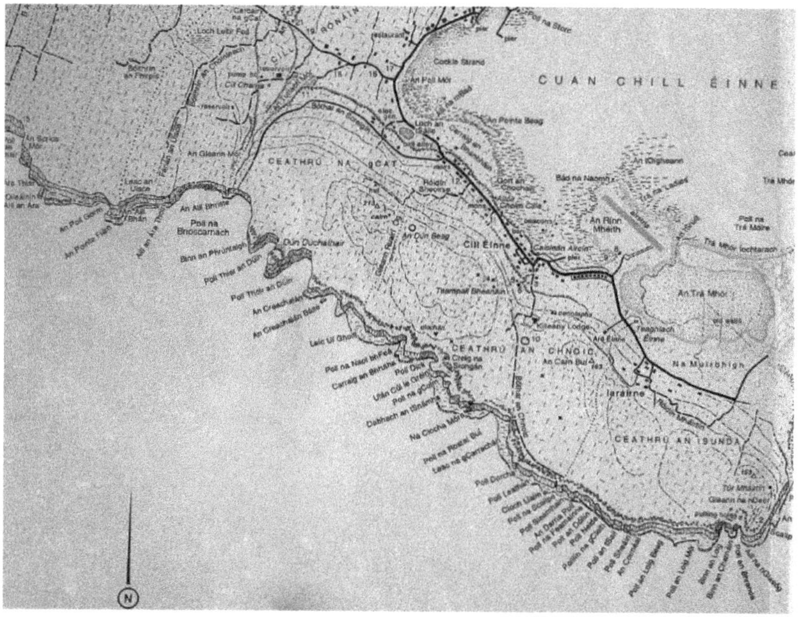

Figure 3. Map detail of south-east coast of Árainn. Folding Landscapes.

Such flexibility, though, also stemmed from the belief that all attempts at mapping are in some small way absurd, "a sustained attempt upon an unattainable goal", and that this nagging ambition towards objectivity might find its peace with something closer to the personal artistic vision ([1], p. 77). At his most personal he even playfully includes an image of his dog on the map where it kills its first rabbit, and of a badger that he stumbles upon in the Burren. As he describes in a later essay on Aran:

> "This horde of men who tramped over the countryside with theodolites and chains so adequately measured its lengths, breadths and heights that I am free to concentrate on that mysterious and neglected fourth dimension of cartography which extends deep into the self of the cartographer." ([1], p. 19).

Recognising the straightforward mathematical survey offered by the Ordnance Survey, Robinson sees his role as populating that Euclidian space with something more psychological and personal. The idea of a personal space, though, also becomes the space for historical, cultural and narrative depth, all of which comes to oppose the international standard of imperial space. The personal becomes an assertion of freedom from that standard which resonates with the island's own freedom from imperial rule.

Whilst these maps are the work of an individual artist, their collaborative element arising from interactions with a broad range of people and their knowledge and memories of the places, meant that Robinson came to view the maps as taking on "aspects of communal creation" ([34], p. 35). *In fact*, it was recently discovered that while writing the content for his first book on Connemara, Robinson would publish his findings in the local newspaper, The Connacht Tribune, and has described the response:

> I had no idea quite how much attention was being paid to them until quite well into the process I found that everyone was waiting for me to turn up. They were quite indignant if I hadn't turned up to them. And they'd have all their information absolutely on the tip of their tongue ready for me. I'd say in a sort of diffident way: "O I'm the man from Roundstone who's making the map", and they'd immediately start "O himself has a stone he wants to show you", "the name of that hill is such and such." ([20], p. 9).

In a range of different ways the "aspects of communal creation" really did involve the communities. The maps became expressions of collective experience, a folding together of discrete forms of knowledge and memory, what David Harvey calls "small heritages", a plural term articulated in contradistinction to the idea of an "Authorised Heritage Discourse", and both the maps and the books together began to constitute a new cultural space that could openly contain really quite diverse variegations ([13], p. 20). The openness of this process found itself very eloquently expressed in a recent exhibition of his Aran map in London. In 2011 Robinson was invited to exhibit in Hans-Ulrich Obrist's "Map Marathon" at the Serpentine Gallery alongside Louise Borgeoise and Ai Weiwei. His contribution was a twenty-two foot vinyl print of his map of the Aran Islands laid on the floor. Come and walk on it, he invited. Come and dance on it. Come and write your name, or your message on it. Pens were provided and people did. The map has gathered all kinds of annotations now, recalling the earlier public participation of his work at Kenwood House but also giving a useful visual metaphor for all those different public contributions that his vision of Aran, the Burren, and Connemara has accrued over the years [35].

4. The Writing: A Quest for Space

"Space", as Robinson describes it, is his preoccupation ([1], p. vi). Aran and Connemara we might more readily and more comfortably describe as *places* but, as was suggested in the introduction, they have become sites in his work for investigating the wider question of space itself as well. Seamus Deane, in a review of the Aran books, suggested that they represented "not perhaps a quest for Aran but a quest to which Aran gives shape and meaning" ([36], p. 9). The nature of space itself is what the wider "quest" has been for in Robinson's work, a quest made all the more intricate by the inward, subjective recesses, the outward, subjective projections, the historical depth, the community feeling, the disciplinary varieties, and the imaginative possibilities, of space in its fullest understanding. The early geometrical spaces that his artwork explored in London, when he was just beginning to invite public and social interaction, found themselves complicated by the two clearly contested and "interlocking" spaces that he uncovered in Aran in tensions between the islands' *dinnseanchas* and the Ordnance Survey maps ([1], p. vi). Space as plural, contested, and yet common, at one and the same time became an experiential reality for him through the map-making and it wasn't long before a growing interest in different, more complex, forms of space began to shoulder out his previous interest in Euclidean conventions.

In the book that completes his *Connemara* trilogy, *Connemara: A Little Gaelic Kingdom*, Robinson addresses precisely this interest, taking, "as a source of metaphor and imagery", the fractal geometry of Benoît Mandelbrot ([37], p. 252). He is prompted to do so by Mandelbrot's 1967 essay "How Long is the Coast of Britain?", which he applies to the intricate folds and convolutions of the Connemara coastline showing that the more closely it is observed the greater the answer until the answer becomes nigh infinite. When he writes of the intricate and changing coastline of this landscape, it is increasingly with a realisation of the inadequacy of Euclidean geometry to represent its complexity. Not only this, but we begin to hear an echo in the language suggesting a parallel with the inadequacy of the Ordnance Survey to represent Irish culture. The land is described as "largely composed of such recalcitrant entities, over which the geometry of Euclid, the fairytale of lines, circles, areas and volumes we are told at school, has no authority" ([37], p. 249). And again, coastlines are "too complicated to be described in terms of classical geometry, which would indeed regard them as broken, confused, tangled, unworthy of the dignity of measure" ([37], p. 249). The lack of "authority" chimes with the book's earlier part on Connemara's histories of political and cultural rebellion and the mention of something "confused, tangled, unworthy of the dignity of measure" could as easily describe the English bafflement and contempt for the Irish *dinnseanchas* as it describes here a mathematical difficulty.

The depths and complexities opened up by fractals can be read also as a way of understanding the manner in which language came to take over the exploration of certain spatial depths that cartography could not capture. The spaces of stories and histories are a part of that "interlocking" of "all our mental and physical trajectories" that he suggests when he claims that "ultimately there is no space but Space", a part of the community's or the culture's intersubjectivity, that empathic belief in a common lifeworld shared by others ([1], p. vi). While few would disagree that our distinctive knowledge, memories, and experiences of space are "interlocking", what Robinson is curious about is the manner in which this "interlocking" takes place, the manner in which it *might potentially* take place, and the role that writing can play in working with the textures of this common but plural "Space". It is in this sense that his interest in "Space" has gone beyond the safeguarding of a body of oral history knowledge precariously balanced on the edge of memory and perhaps that his work has come to exceed and extend the work of deep mapping as well as anticipate it. At least, it draws attention to space as a social and artistic medium rather than as an area or volume in such a way that may be useful for deep mapping as it develops its practice.

In a recent essay he tells us that for a good many years now he has been building a computer database on CD-ROM of all the topographical knowledge he has been collecting, a way of indexing and preserving the research in a more detailed manner than the paper maps allow and in a more systematic manner than the literary books allow. In this essay—called "The Seanachaí and the Database", where "*seanachaí*" comes from the same root as *dinnseanchas*—he begins to weigh the strengths and weakness of his database against the strengths and weaknesses of the *dinnseanchai* (or "keeper of topographical knowledge and lore of place") ([38], p. 46). He finds that the database "transcends" the local inhabitant's mental gazetteer "in powers of recall and logical organization"; it is searchable and it has no limit to the amount of information it can store ([38], p. 47). However, the database falls far short when it comes to "ambiguous or doubtful data" and "as a memorandum of lifelong inhabitation" ([38], p. 47). The *dinnseanchaí* is not simply a vessel containing historical and cultural information but perceives and creates meaningful relationships between the different parts of the retained lore; he or she is capable of ordering or reordering the history and lore according to values related to that lifelong inhabitation. This is echoed in Ian Biggs' claim that deep mapping's preoccupation with bringing the past to light always has an eye on the contemporary as well and on necessary investigations and productions of meaning "so as to enact constructive reconciliations in the present" ([9], p. 5). Though Robinson does not make such a claim himself, it is precisely this lively negotiation of relationships and the capacity to create meanings that we see in his writing, and that distinguishes it from the maps and the database.

The capacity of his prose to draw on detailed knowledge of, for example, botany, archaeology, folklore, geology, and history, all in the same chapter or essay—his attempt at "interweaving more than two or three at a time of the millions of modes of relating to a place"—shows something more deliberate and creative than the bringing forth of peripheral and precariously located knowledge that we find in the maps and their accompanying booklets ([2], p. 363). It shows an attempt to make the discrete layers of space belonging to different disciplinary perspectives known to one another and present to one another for the reader. There is a creative practice in this, reaching into the imagination of the artist at one extent and into the cultures with which a given place is invested at the other. This gives rise to two identifiable traits in his written work. Firstly, as Pippa Marland has shown, there is an extraordinary range of experimental writing styles through which he moves, self-reflexive moments of pastiche and parody, mischief and humour, moments of elaborate construction suddenly undermined by irrepressible self-doubt, all of which engage with the often overwhelming possibilities available to an author seeking to do a kind of literary justice to place ([8], pp. 19–21). This variety of registers is something that Susan Naramore Maher describes as characteristic of the form of writing associated with deep mapping in the United States, and something that she reads through Mikhail Bahktin's writings on heteroglossia and the dialogic imagination in William Least Heat-Moon ([39], p. 7). However, secondly, there is a more consistent and identifiable formal trait in Robinson's recurrent endeavour to relate and intertwine distinctive perspectives on place, looking for what he calls, after E. O. Wilson, "consilience" ([20], p. 10). The abiding question that he puts to the test again and again is "can the act of writing hold such disparate materials in coexistence?" ([2], p. 210).

One such example of this we have in *Connemara: The Last Pool of Darkness* in which he describes the large cleft in the hillside that forms the valley of Little Killary near the coast. Here in the land's unusual geology "ancient uncouth states of the earth have been broken through and thrust one over the other" ([40], p. 2), then gouged and worn away by glaciers. At the head of the valley there is a chapel and well dedicated to the little known St Roc where people used to bring their dead for funeral rites. Local folklore explains the dramatic landform by suggesting that it was here that St Roc struggled with the Devil: as the Devil tried to pull him away to Hell on a chain, St Roc resisted "so violently that the chain cut deep into the hillside, creating the pass and funerary way" ([40], p. 2). "Thus geology reveals itself as mythology," Robinson claims. However, it was also on the edge of Little Killary that Wittgenstein once stayed during a period of retreat from Cambridge while struggling with the philosophical argument about "the difference between seeing something, and seeing it as something" (the famous example is of the shape that appears as a duck from one angle and a rabbit from another) ([40], p. 1). This particular branch of Wittgenstein's thought has huge significance for Robinson insofar as different perceptions can give way to different explanations of place residing in the same "Space" ("there are more places within a forest, among the galaxies or on a Connemara seashore, than the geometry of common sense allows" ([37], p. 252)). So Robinson suggests: "In some future legendary reconstitution of the past it will be Wittgenstein's wrestling with the demons of philosophy that tears the landscape of Connemara" ([40], pp. 2–3). Here, mythology, geology, and philosophy are all brought into a resonant proximity by geographical and historical association and intertwined through the narrative of the essay. Wittgenstein's own problem about seeing something as something else is playfully deployed and perhaps even celebrated by revealing a literary form of the duck/rabbit diagram in the form of geological rift/St Roc's struggle/Wittgenstein's struggle. It is this prose trait that recurs throughout the writing in different forms and that Robinson describes as a form of "consilience".

For E. O. Wilson, consilience is a means of bringing the, predominantly scientific, disciplines together in the joint endeavour of expanding the horizons of human knowledge but it too has a relationship to the religious past for him:

> We are obliged by the deepest drives of the human spirit to make ourselves more than animated dust, and we must have a story to tell about where we came from, and why we are here. Could Holy Writ be just the first liberate attempt to explain the universe and make ourselves significant within it? Perhaps science is a continuation on new and better-tested ground to attain the same end […] Preferring a search for objective reality over revelation is another way of satisfying religious hunger. ([41], p. 5).

There is, however, an important difference here. Wilson's understanding of consilience is fundamentally teleological with religion fumbling in the dark behind us and the light of scientific explanation on "better-tested ground" ahead of us (theologians might find themselves irked by the thought that religion is a naïve form of scientific endeavour). However, Robinson's understanding of consilience is not so teleological and is, instead, interested in the thoughts that appear as different layers of knowledge coincide, as different ways of seeing the same thing "as something" multiply its phenomenological and intersubjective possibilities. Robinson thinks like an artist, Wilson like a scientist.

As Wilson himself suggests, somewhat reductively, "the love of complexity without reductionism makes art; the love of complexity with reductionism makes science" ([41], p. 54). Michael Quigley has also suggested that "no book containing such a vast amount of detail on such a small portion of landscape could possibly be sustained were it not for its intrinsic literary quality" ([42], p. 117). The artist and the cartographer eventually find a curious fulfilment of their quest for richer and richer forms of space in the contours and possibilities made available through language and the literary imagination.

Robinson has described the "base-triangle" of his philosophy of space as "that formed by the three church-towers of Proust's Martinville". ([1], p. 19) The "base-triangle", in cartography, is the measure from which all other measures are unfolded, one triangle after another. For the Ordnance Survey this first base-triangle was a precise measure taken with extraordinary care over a two-month period with help from members of the Royal Artillery on Hounslow Heath in the summer of 1791 ([32], pp. 124–26). From that measure the survey worked outwards until it had taken in the whole of the British Isles. For Robinson, the reference to Proust suggests something much more deeply felt and subjective that is related to the impulse to write. The mysterious intensity of feeling that Proust describes upon watching from a coach window the triangulation of Martinville's two church towers with Vieuxvicq's one ignites in his narrator the need to write a response. The sense "that something more lay behind that mobility, that luminosity, something which they seemed at once to contain and to conceal' wrenches open an irresistible need to respond creatively which, once fulfilled, produces feelings of extraordinary elation ([43], p. 184). Robinson's suggestion, then, that this moment in Proust serves as his own "base-triangle" expresses a very primal and mysterious response to what a place might at once "contain and conceal" then (and we might think back to the image of the piper in the cave again here). It alludes, again, to an immanent metaphysics. Part of what a place contains and conceals for Robinson is, of course, these "interlocking" spaces both of historical depth and diverse subjective and disciplinary perspectives. It contains these, but it also contains the endlessly deferred promise of a total form of spatial revelation, a revelation implicit in the very idea of concealment itself.

The central philosophy, introduced in the first Aran book, but running through both of the Aran books, is Robinson's idea of "the good step", an ideal, single footfall that traverses a portion of the earth while somehow containing an unthinkable awareness of all possible ways of knowing the place it is traversing, what he describes as an "unsummable totality of human perspectives" ([2], p. 8). The "good step" is an aspiration to do a form of cognitive and spatial justice to a given place, to unlock it spatially. It is an ideal realisation, if not revelation, of all that is "contained and concealed". It is, of course, he declares, "inconceivable" in the end but this does not prevent it, as an ideal, from shaping his attitude to the next step *ad infinitum*, honouring the impulse to reach for and respond to the contained and the concealed even if the revelation may only ever be partial ([2], p. 363). What is particularly interesting about this process of always partial revelation is the creative and imaginative work that is required to bring it about, particularly the way that it is the creative imagination of the prose essay and its aesthetic of consilience that brings the different spatial layers into productive contact with one another for the reader. It is in this sense that space takes on the qualities of a social and artistic medium, one worked over by language; worked in the sense that it is *produced*, but with an effortlessness that gives it the impression of *revealing* its own nature instead. It is through this interesting overlap between the production of space and the revelation of space that we learn something about the work of deep mapping by reflecting on Robinson's career. Deep mapping explores a means of place-based social transformation but Robinson's practice shows that this can happen fruitfully through creative and inclusive work at the level of space, when space is privileged as something with a psychological texture, a social fabric, and a cultural value, as a medium invested with and constituted by multiple perspectives at the intersection of history and community.

5. Conclusions

Robinson's secular metaphysics of "Space" are, in the end, a deep complement to his investment in place. They are, first, his route *to* place, out of the isolated abstraction of twentieth-century metropolitan modernism. They were always what place was embedded within cosmically for him, but they became the lifeblood of place too, its interior labyrinth through which contested versions of the same place were found to be "interlocking". It is in the latter of these understandings of space that he came to realise the potential for the positive social contribution of his work, producing and developing deep layers of space in socially engaged forms of *ad hoc* heritage work, *ad hoc* insofar as they were generated from the ground up, outside of any institutionalised heritage discourse. Recent work in critical heritage studies has argued for the recognition of the fundamentally creative nature of all heritage work, that even when it appears to be simply preserving the past from the threat of contemporary life, it is nonetheless producing a narrative for the present and into the future ([44], p. 100). From the mid-1980s there have been calls for contemporary artists to be more involved in heritage work for the way that they "have continued to struggle with the material of the present" when the heritage industry has turned its back on it ([45], p. 144). Robinson's work in Ireland represents just this kind of relationship between heritage and contemporary art. While his early map-making revealed those "interlocking" spaces, his later written work began to interweave them in search of that "consilience" that could produce and reveal space simultaneously. In this sense the space of the writing returns full circle to the abstract and gestural work of the earlier artist, but this time the space being worked is not merely geometrical but rather bears and embodies place too in all its fractal minutia as it does so. In this sense Robinson writes as he mapped and as he painted, as an artist whose distinctive investigation, practice, and aesthetic of space reveals space itself to be a medium approached by various means.

Place and social engagement are ideas that might be thought antithetical to modernism, and Robinson's work does show a disruptive tension that emerged in the 1970s. However, there is another way of thinking about this. Recent work on "regional modernisms" [46] and "archipelagic modernism" [47] suggests that, as much as the equation between modernism and the city is being rethought historically, its legacy might also be traced in artists like Robinson and perhaps even in the wider project of deep mapping itself. As such, Robinson offers a very singular and surprising view on the relationship between modernism and deep mapping, a curious line of inheritance, if not a bridge. As much as deep mapping is about history, community, and place, it too is concerned with the imaginative manipulation of space through innovative and experimental practices in an antagonistic relationship with modernity. Place is increasingly becoming the site of such experimental practices in the twenty-first century, revealing itself to be as fissured and fractured, as protean and volatile, as *deep* in its cavernous passages as the mind ever was for modernism at the beginning of the twentieth century. Robinson's work draws attention to the mechanics of place heritage as an imaginative social and spatial work, as what practitioners such as Ian Biggs are increasingly revealing to be perhaps *the art of place* itself.

Acknowledgments

I would like to thank members of the Atlantic Archipelagos Research Consortium with whom I worked on the first public conference to discuss Tim Robinson's work in Galway in 2011 (funded by the British Academy). I would also like to thank Andrew McNeillie who first encouraged me to read Tim Robinson's work many years ago.

Conflicts of Interest

The author declares no conflict of interest.

References

1. Tim Robinson. *Setting Foot on the Shores of Connemara and Other Writings*. Dublin: The Lilliput Press, 1996.
2. Tim Robinson. *Stones of Aran: Pilgrimage*. London: Faber and Faber, 2008.
3. Edward S. Casey. *Getting Back into Place*. Bloomington: Indiana University Press, 1993.
4. John Elder. "Introduction: Unfolding the Map." In *Connemara and Elsewhere*. Edited by Jane Conroy. Dublin: Prism, 2014, pp. 1–25.
5. Tim Robinson. *My Time in Space*. Dublin: The Lilliput Press, 2001.
6. Timothy Drever. "Untitled." 1969. Available online: http://www.tate.org.uk/art/artworks/drever-untitled-p04228 (accessed on 15 July 2015).
7. John Wylie. "Dwelling and Displacement: Tim Robinson and the Questions of Landscape." *Cultural Geographies* 19 (2012): 365–83.
8. Pippa Marland. "The 'Good Step' and Dwelling in Tim Robinson's Stones of Aran: The Advent of 'Psycho-archipelagraphy." *Ecozon@* 6 (2015): 1–6.
9. Ian Biggs. "'Deep Mapping': A Brief Introduction." In *Mapping Spectral Traces*. Edited by Karen Till. Blackburg: Virginia Tech College of Architecture and Urban Studies, 2010, pp. 5–8.
10. Ian Biggs. "The Spaces of 'Deep Mapping': A Partial Account." *Journal of Arts and Communities* 2 (2011): 5–25.
11. Mike Pearson, and Michael Shanks. *Theatre/Archaeology*. London: Routledge, 2001.
12. William Least Heat-Moon. *PrairyErth (A Deep Map): An Epic History of the Tallgrass Prarie Country*. New York: Mariner Books, 1991.
13. David C. Harvey. "The History of Heritage." In *The Ashgate Research Companion to Heritage and Identity*. Edited by Brian Graham and Peter Howard. London: Ashgate, 2008, pp. 19–36.
14. Anonymous. "Folding Landscapes." *Geographical Magazine* 61 (1988): 60.

15. Tim Robinson. *The View from the Horizon*. London: Coracle Press, 1999.
16. Clement Greenberg. "Avant-Garde and Kitsche." In *Art in Theory: 1900–1990*. Edited by Charles Harrison and Paul Wood. Oxford: Blackwell, 1999.
17. Timothy Drever, Ed Herring, Peter Joseph, and David Parsons. *Survey 69 New Space*. Camden: Libraries, Arts and Amenities Committee, 1969.
18. Timothy Drever, and Peter Joseph. "Outside the Gallery System: Two Projects for Kenwood." *Studio International* 177 (1969): 255.
19. Alan Sinfield. *Literature, Politics and Culture in Postwar Britain*. London: Continuum, 2005.
20. Jos Smith. "A Step towards the Earth: Interview with Tim Robinson." *Politics of Place* 1 (2013): 3–11.
21. Stuart Brisley. "Environments." *Studio International* 177 (1969): 266–69.
22. Clarrie Wallis. "Making Tracks." In *Richard Long: Heaven and Earth*. Edited by Clarrie Wallis. London: Tate Publishing, 2009, pp. 33–61.
23. Raymond Williams. *The Politics of Modernism: Against the New Conformists*. London: Verso, 2007.
24. Nicolas Grene. "Introduction." In *Travelling Ireland: Essays 1898–1908*. Dublin: The Lilliput Press, 2011.
25. J. M. Synge. *The Aran Islands*. London: Penguin, 1992.
26. Colin Kirkpatrick. "Richard Long: No Where." *Sperone Westwater*. Available Online: http:\\www.speronewestwater.com (accessed on 14 October 2012).
27. Tim Robinson. *Connemara: Listening to the Wind*. London: Penguin, 2007.
28. Jennifer Westwood, and Jacqueline Simpson. *The Lore of the Land*. London: Penguin, 2005.
29. Michel De Certeau. *The Practice of Everyday Life*. Translated by Steven Rendall. Berkeley: University of California Press, 1984.
30. Charles Bowen. "The Historical Inventory of the Dindshenchas." *Studia Celtica* 10 (1975–1976): 113–37.
31. J. H. Andrews. "Paper Landscapes: Mapping Ireland's Physical Geography." In *Nature in Ireland*. Edited by John Wilson Foster. Dublin: The Lilliput Press, 1997.
32. Rachel Hewitt. *Map of a Nation: A Biography of the Ordnance Survey*. London: Granta, 2010.
33. Patrick Curry. "Elegies Unawares." *Times Literary Supplement* 8 (1995): 13.
34. Brian Dillon. "An Interview with Tim Robinson." *Field Day Review* 3 (2007): 32–41.
35. Anonymous. "Map Marathon." *The Serpentine Gallery*, 2010. Available online: http//www.serpentinegallery.org. (accessed on 19 October 2012).
36. Seamus Deane. "Ultimate Place." *London Review of Books* 11 (1989): 9.
37. Tim Robinson. *Connemara: A Little Gaelic Kingdom*. London: Penguin, 2011.
38. Tim Robinson. "The Seanchaí and the Database." *Irish Pages* 2 (2003): 43–53.
39. Susan Naramore Maher. "Deep Mapping the Great Plains: Surveying the Literary Cartography of Place." *Western American Literature* 36 (2001): 4–24.
40. Tim Robinson. *Connemara: The Last Pool of Darkness*. London: Penguin, 2009.
41. E. O. Wilson. *Consilience*. London: Abacus, 1999.

42. Michael Quigley. "Natural History and History of Ireland." *Irish Historical Studies* 31 (1998): 115–23.
43. Marcel Proust. *Remembrance of Things Past, Volume One: Swan's Way*. Translated by C. K. Scott Moncrieff. New York: Henry Holt and Co., 1922.
44. Rodney Harrison. *Understanding the Politics of Heritage*. Manchester: Manchester University Press, 2010.
45. Robert Hewison. *The Heritage Industry*. London: Methuen, 1987.
46. Neal Alexander, and James Moran, eds. *Regional Modernisms*. Liverpool: Liverpool University Press, 2013.
47. John Brannigan. *Archipelagic Modernism*. Edinburgh: Edinburgh University Press, 2014.

A Strange Cartography: Leylines, Landscape and "Deep Mapping" in the Works of Alfred Watkins

James Thurgill

Abstract: In 1921 the photographer, antiquarian and amateur archaeologist Alfred Watkins, delivered his newly formed thesis on the origins of ancient alignments in the west of England to the Woolhope Naturalists' Field Club of Hereford. Watkins posited a correlation between ancient forts, moats, mounds, churches, trees and place names, which he had shown to produce straight lines running across the landscape. In 1922 Watkins published his first book on the subject, *Early British Trackways*, mixing amateur archaeology, social history and supposition to introduce what Watkins named "leylines" and setting out the guidelines for other would-be ley hunters. This paper explores Watkins' ley hunting as a practice of "deep mapping", examining its use as an applied spatial engagement with the hidden trajectories of the landscape. In addition to providing a concise cultural history of the leyline, with particular reference to the works of Alfred Watkins, this paper develops a critical engagement with ley-walking through an auto-ethnographic response to a leyline that has been mapped and walked in Norfolk, England.

Reprinted from *Humanities*. Cite as: Thurgill, J. A Strange Cartography: Leylines, Landscape and "Deep Mapping" in the Works of Alfred Watkins. *Humanities* **2015**, *4*, 637–652.

1. Introduction

This paper investigates spatial engagements with the landscape through the spiritual-cartographic concept of leylines. The identification and walking of leylines will be discussed as a method of "deep mapping", a manner in which to "record and represent the grain and patina of place" ([1], pp. 64–65). Leylines work to uncover and trace occulted trajectories, believed (by some) to form a network of ancient trackways across the landscape, interconnecting with points of both historical and sacred importance. This article aims to highlight the importance of the leyline as a tradition of deep mapping as well as demonstrating the wider significance of the ley in the development of the spatial humanities and indeed, the experimental geographical methods that might be used to engage with leylines.

Leylines make for a rather unfamiliar cartographic system, one that provides a method for navigating one area of the country to another purely by walking along straight lines, indicated by a series of alignments in both naturally occurring and manmade landmarks. In addition to providing a brief history of the leyline, with particular reference to the work of Alfred Watkins, I will develop a critical engagement with leylines, providing an auto-ethnographic response to the walking of a ley. Through a personal account of identifying and walking a ley, the auto-ethnography will be used to illustrate the rich experiential qualities of leylines within the practice of deep mapping.

There has been a substantial interest in occult spatiality in recent years [2–19]. Building upon the existing catalogue of research in the area of the spatial and the occult, this research will broaden the foci of the "unpopular geographies" of spectrality and monstrousness to include what I refer to as deeply topological belief systems, bringing original case study material into the scope of *strange* affectual landscapes.

In this paper, I will articulate a discourse of place that explicitly deals with the relationality that is expressed through the spatial practice of ley hunting and which allows for both history, memory and narrative to be seen linearly: mobile and navigable. Strictly speaking, then, this is to be a discussion of cartographics, of maps and of mapping, albeit in an arcane fashion. Furthermore this is an enquiry, in the broader sense, into how lines interlope into the realms of archaeology, spirituality and placial ontology. Finally, it considers how lines might indicate particular trajectories that lend themselves to a re-imagining of or re-acquainting with our surroundings.

2. Alfred Watkins: The Ley of the Land

In the September of 1921, 66 year old Alfred Watkins delivered a talk to the Woolhope Naturalist's Field Club presenting his thesis that the landscape could be mapped as a series of straight lines or "alignments" that "connected ancient burial mounds, monuments, barrows, ditches, castles, ponds and trackways" ([20], p. 81). Watkins went on to name such alignments "leylines".

Already well published in the fields of photography and local history, Watkins' long walks across the undulating Herefordshire countryside had generated a huge amount of data, both written and photographic; his work indicated the age and locations of ancient monuments, hill forts, pre-Roman villages, churches, trees, follies and more. Watkins noticed that these objects and their places appeared to be aligned in an unmistakably intentional fashion: long straight trackways appeared to etch their way across the landscape and Watkins attributed the causation of these lines as matching the age of the monuments that marked them: early man. Watkins went on to describe the conception:

> Imagine a fairy chain stretched from mountain peak to mountain peak, so far as the eye could reach, and paid out until it touched the high places of the earth at a number of ridges, banks and knowls. Then visualise a mound, circular earthwork, or clump of trees, planted on these high points, and in low points in the valley, other mounds ringed round with water to be seen from a distance. Then great standing stones brought to mark the way at intervals, and on a bank leading up to a mountain ridge or down to a ford the track cut deep so as to form a guiding notch on the skyline as you come up. In a bwlch or mountain pass the road cut deeply to show as a notch afar off. Here and there, and at two ends of the way, a beacon fire used to lay out the track. With ponds dug on the line or streams banked up into "flashes" to form reflecting points on the beacon track so that it might be checked when at least once a year the beacon was fired on the traditional day. All these work exactly on the (ley) line ([21], p. 218).

According to Watkins what he was describing was not, then, an enchanted genealogy, but rather simply a diagnosis of the facts as they presented themselves to him. And in fact his diagnosis had some precedents: similar conceptions of extensive trackways had previously been recorded, for example by Hilaire Belloc, who had posited and indeed mapped semi-ancient pilgrim paths in his The Old Road [22]. Belloc displayed a deep sense of connection to trackways, remarking that the road "gives a unity to all that has arisen on its way" ([22], p. 4). But although Belloc had made early claims for a series of forgotten travel routes, he had limited his scope to the movement of early pilgrims: the idea of an interrelated network of tracks, as later seen in the work of Watkins, had never occurred to him.

It was not the idea of the leyline that formed Watkins greatest challenge. He had derived the concept of the ley from the data that he had accumulated quite easily. Rather, it was formulating the theory into something solid, discovering a way to convey his discovery to a wider audience. He finally achieved this solid explanation in the form of an oral paper delivered to his archaeological peers at the Woolhope Naturalist's Fieldclub in 1921 [23]. The first paper was entitled Early British trackways, moats mounds camps and sites. In this initial formulation, Watkins developed his belief that there were a significant number of megaliths and monuments that fell into almost perfectly straight lines within the vicinity of his home county of Herefordshire. He provided hand drawn maps and vast quantities of field research to account for this belief in an intentional linear correlation. The term ley developed as a result of the number of alignments crossing through places whose names ended with "ley" [21]. Watkins believed there to be a functional reasoning for the existence of these sightlines, their purpose being chiefly to guide the people of early Britain through the landscape. This he posited as being the predominant reason that the alignments had arisen; in the absence of cartography, early communities had required some indicators that they were travelling along the correct route. Watkins' aim in this early paper was to establish the initial development of "a framework for a new knowledge" ([23], p. 34), a new method through which to understand the placement of "things" in the landscape.

Unfortunately there is no record of the reception of Watkins' paper by the Woolhope Club, Shoesmith [24] notes that the Club did not record any commentary of the presentation in their minutes, though some evidence of Watkins' continued output with the organisation can be found in later publications of the Club's proceedings. Whatever the reaction, Watkins set out to publish his findings and produced his first book on the subject in 1922: a publication of the original lecture to the Woolhope in its entirety, complete with accompanying monochrome plates, released a year after he gave his talk. In the foreword to the text, addressed to "the average reader", Watkins launches a diatribe against his contemporary antiquarians, who he accused of having "alternated between a misty appreciation of hill-tracks and ridgeways, and an implied depreciation of all track makers before the Romans came" ([23], p. 7). The book was generally well received by the general public, and Watkins is known to have kept a scrapbook of newspaper cuttings documenting the publicity that his ley theory had attracted. He also continued to publish his research in the proceedings of the Woolhope Club [25].

Keen to have his geometric patterns recognised, Watkins produced his most significant and widely received work, The Old Straight Track in 1925. Where Early British Trackways merely sketched the initial theory, in The Old Straight Track Watkins produced a set of evidence based reports of his walks along these hidden linear routes; this included sketching of ley trajectories, extensive descriptions of the types of indicators of alignments and further photographic verification. Watkins explained his theory in detail over two hundred and twenty six pages, making this book his most substantial and articulated output on the subject of leylines. "The Old Straight

Track" [21] as Watkins termed the leyline, was to be seen across the country and the richly illustrated and meticulous record he had produced presented others with the chance to go out and explore the countryside looking for forgotten ancient pathways by themselves. The positive reception of the second book's reception became the impetus for Watkins writing a third and final book on the subject of leys. Acting as something of a training guide for fellow ley enthusiasts, The Ley Hunter's Manual: A Guide to Early Tracks was completed by Watkins in 1927 (mass published in 1977 [26]), and gave clear instructions on how to identify and uncover leys using maps and also at sight.

For Watkins, leys indicate a forgotten and ancient knowledge, or rather a composite of knowledge; lost, historic and immanent. Watkins insisted that such an understanding was embedded within the landscape, left for us to uncover:

> Knowledge is only to be gleaned from three types of evidence. Firstly and chiefly from what exists or is recorded on or in the earth of the work or remains of man of that period. Secondly, from what can be gleaned and surmised in place-names and words, for it is often forgotten that words were spoken in Britain for more centuries before they were written down than there have been centuries of written record, and there are indications that many words elements come down through both periods. Thirdly, from folk-lore legends; lingering fragments of fact disguised by an overlay of generations of imaginings ([21], p. xix).

In its Watkinsian form then, ley hunting is a way of retracing history: uncovering site narratives and reimagining the landscape, it is a particular way of knowing place, of developing a biography for and of place.

Paul Devereux [27] notes that Watkins' leylines marked specific path or trackways across the country that joined settlements with sites of ritual and sites of trade. In other words, what Watkins uncovered was a map of knowledge engrained into the landscape, with its paths forming clear nodes at sites of importance, a systematic approach to a navigating of the countryside, or as Matless describes it "a specific topographical geometry" ([20], p. 83), a topographical geometry consisting of traces, of additive lines. Indeed, the very idea of the line suggests movement, or at least the opportunity for mobility. A line has two connected points; it runs from and between one and the other. In conjecturing that "Life on the spot surely cannot yield an experience of place, of being somewhere" ([28], p. 2), Ingold sets up a theoretical trajectory which supports a linear spatiality of place, an experience of being "somewhere" as reliant upon movement. In concluding that "(t)o be a place, every somewhere must lie on one or several paths of movement to and from places elsewhere" [28], Ingold suggests that places can only exist as places if they are in connection with one another, in dialogue with other places. Ingold [28] goes on to classify two manifestations of the linear form; threads, an entanglement of lines upon a surface, and traces that can either add or remove from a surface linearly.

Leys belong to that classification where lines are engendered through a process of production; they trace connections between and bind ancient objects to the landscape, and as such, are valorised by something far more than just the provision of scale, distance, direction or destination. As lines, leys provide meaning; they suggest that our ancestors knew the land around them in a manner that many believe has been lost in the contemporary urban existence. Leylines enrich history through a deepening of our interrelation with the natural world and of greater import still, they provide a way of getting back into the landscape, of mapping, retracing, reimaging and resurrecting our position within and amongst places. Leylines imply a meaning between places; they indicate a movement between spheres, areas that might once have appeared as the hostile milieu of otherwise abstract space. Watkins' leys, then, can be seen to make sense of things, to provide clarity through the making of lines and a close or "deep" mapping of landscape. Stephen Daniels explains Watkins' thesis not so much as a theory but as "a sorting out of previously unrelated and unnoticed information 'embedded' in the mind and on the ground" ([29], p. 6). Daniels may be right, but the ramifications of this "sorting out" have remained unexplored almost ninety years on. To be sure, when discovering leys, we are unveiling our surroundings as a "topography of generosity"; a space that creates and offers us a multitude of places in its very unfolding, providing new routes into and across the landscape. Through leylines, the landscape affords us possibilities for the (deep) mapping and reimagining of the world around us, ones that might not ordinarily have come to our attention.

If one is to follow Watkins' line of thought then stones, mounds, ponds, trees, beacons, churches, camps, and hillside notches all correlate along natural trajectories throughout the landscape along which early man would have ventured. The idea of the ley was, for Watkins, as we have seen, quite clearly a scientific one; however although he himself steered clear of the occult, John Michell's investigation of geomancy suggests that Watkins "remained aware that there was a far deeper significance to the leylines than he had given mention to through the supposition of ancient trackways and trade routes" ([30], p. 189). Naturally, the genesis of such a novel idea—as it had been perceived at the time—left itself wide open for various interpretations; mysticism quite quickly became one of the frameworks in which leys would be viewed and, forsaking the intentions of Watkins, is now probably the trope with which leylines are most commonly equated.

3. Leylines after Watkins

Another notable figure in the development of leylines was Guy Underwood, who published "The patterns of the Past" [31] in 1968. Like Watkins, Underwood too believed in a hidden order of things and conjectured that, despite the beliefs of the academy, there was surmountable evidence of leys across Britain, most notably at Stonehenge. Contra Watkins, Underwood's interest in alignments lay not in trackways per se, but rather in the formation of pathways and patterns by water, a phenomenon he termed the "Geodetic line" [31]. Underwood categorised Geodetics as existing in three kinds of line: the "water line", the "aquastat" and the "track line":

> The three "primary lines" are so called to distinguish them from the secondary effects they produce. The water line, the aquastat and the track line have much in common: they appear to be generated within the Earth; to involve wave-motion; to have great penetrative power; to form a network on the face of the Earth; to affect the germination and manner of growth of certain trees and plants; to be perceived and used by animals;

> to affect opposite sides of the animal body, and to form spiral patterns. They are controlled by mathematical laws which involve in their construction the number 3; and in their spiral patterns, the number 7. They played a prominent, and possibly fundamental, part in the religion of many widely scattered primitive peoples ([31], p. 34).

Not only did Underwood distinguish his lines from Watkins leys in terms of their causation, he also made far more obvious links with the occult and divination. Connections with numbers, Earth mysteries and an ascribed power of intent are all discussed with direct reference to leys in Underwood's work, which as a result leads to the mystification of leys. Following an occult trajectory does more for Underwood than just sensationalising the already existing mystery of landscape patternation; it offers an unprecedented agency to both the Earth and the leyline, and thus consequently emancipates leylines from the belief that they were constructed by humans. Such a conjecture is what formed the basis for the migration of leylines from archaeology into mysticism. By positing that an unseen "magic" was at work in forming the landscape, Underwood created a platform in which to use leys. Underwood posited that he was able to work with leylines via a harnessing of the Earth's energies, employing supposedly ancient practices such as that of dowsing, a tradition that I shall discuss in further detail later in this paper.

In fact The Patterns of the Past is far more of a guidebook to using ley-focused practices (in their various incarnations) than a reimagining of the historical ponderings of Watkins. Together with other ley enthusiasts at the time, Underwood helped to give rise to archeoastronomy; the study of the connectivities between sacred monuments and the celestial. Leylines became something of "a mode for realising enchantment" ([32], p. 23). When combined with theories of mysticism, leys provide an underpinning to the essentialism of place and the initiating of a cartography of enchantment.

In their mystical guise, as presented by underwood, leylines seek to map the sacred as a series of interconnected, interdependent sites of enchantment which themselves dwell-in-the-world. To this end, so-called New Agers, followers of earth mysteries, writers, and later, psychogeographers and artists, all gained a new way of connecting the Earth to mysticism. Leylines became perceived as a network of supernatural energies superimposed upon the surface of the Earth. This strange hybrid of geometry, naturalism and mysticism is known as geomancy, a term defined by Graham Harvey, to describe the discerning of flows of energy that are creative and supportive of particular forms and mapping them in a linear manner across the landscape [33].

The wider influence of Watkins' leanings on landscape or field archaeology has been significant for the development of a number of place-based theories relating arcane knowledge of the Earth to our early ancestors. Indeed, the turn toward the mysteries of the prehistoric past seen during the "Celtic Renaissance" of the 1960s and 1970s led to what John Michell described as "megalithomania" ([34], p. 5). Much of the work derived from Watkins that developed within this period, for example the archaeo-exploratory work of John Foster-Forbes [35], came across as far more esoterically inflicted than archaeological. Together with the assistance of the psychic, Iris Campbell, Foster-Forbes built upon the more speculative aspects of Watkins' work to conduct field studies of ancient sites across Britain. Foster-Forbes based approach on the idea that megalithic sites were important to the ceremonial cultures of early British people, but then entwined the history of the creation of these spaces with a fantastical history, positing their construction as the result of giants.

Like Foster-Forbes, T.C. Lethbridge [36,37] also took a more mystical approach to Watkins' work, looking to correlations between mythology and the archaeological to provide answers as to why ancient sites existed and how they were connected. Similar sentiments can be found in the works of Pennick [38], Pennick and Devereux [39], Devereux [27,40,41], Cope [42] and Sullivan [43], with each conjecturing an underlying mystical energy to the ley. Leylines are thus understood and studied in two different ways. On the one hand there is an evidence-based application to the understanding of alignments through a mapping and walking of their routes; on the other, there is the belief that leys relate to an intrinsically enchanted and lost past hinting at a mystical "Golden Age" of human existence. Such a position resonates in Cope's claim that our land "once undulated with the all that was the wonder of Mother Earth" ([42], p. ix). This growing trend towards a spiritual inflection of the prehistoric landscape brought with it a revival of folk traditions and land divination, parts of which were employed by geomancers like Underwood in order to commune with the resonating energies of the earth [31].

Dowsers and Diviners

The practices of divining or "dowsing" the landscape show little fidelity to the Watkinsian origins of leylines, but remain widely used as a method for tracking linear land patterns. The practice of dowsing is in itself a spatial engagement, utilising material objects in order to locate and navigate unseen energies in the natural environment. In an online introduction to dowsing, the British Society of Dowsers (BSD) describe the practice of dowsing as being "to search with the aid of simple handheld tools or instruments, for that which is otherwise hidden from view or knowledge" [44]. The BSD were formed in 1933 by Colonel A. H. Bell, who founded the society in order to promote the profile of dowsing throughout the British Isles [44]. "Dowsing", continues the BSD, is foremost the "art of discovering the presence of energies, substances, objects or missing persons or things not apparent to the senses, usually by using rods or pendulums" [44]. The rods described by the BSD come in various forms; L shaped rods, V-rods or tools fashioned from willow or hazel. The rods or pendulum are employed in such a way that they demand mobility from the dowser. As the dowser walks the landscape in search of a leyline the tool responds; it moves as of its own accord, responding to the energies or geodetics resonances of the Earth: the power of the line. In this sense, sacrality can be physically traced: the material objects (rods) indicate the line and the experience is rendered (phenomenologically) authentic through the embodied practice of the dowser moving within the landscape, of living the mapping process first hand.

Dowsing practice can also be employed in a similar manner to the Ouija or divination board, with some dowsers posing questions to the spirit(s) of the place, its *genius loci*, and allowing the rods to act as a mediator, moving in a predetermined pattern in order to answer "yes" or "no". In his *Ghosts and Divining Rod*, Lethbridge [37] notes the similarities between the dowsing of Earth energies and that of spirit energies, claiming that either phenomenon could be posited as types of recordings to which the diving rods are in tune. Wilson [45] follows on in this rather more mystical trajectory with his positioning of dowsing as a method that works to reconcile man with Earth.

Whilst Watkins himself never used dowsing rods in his experiences with leys, this has not stopped the more spiritually minded followers of ley theory from putting such practices to use. Dowsers and ley hunters work to various methodologies and belief systems: spiritual, water sourcing, energy tracing, archaeological; each employs a different set of tools. Accessibility, with regard to employing such practices, is unrestricted—providing the land is designated for public use or in the case of private landownership, permission has been granted. The ley hunter or dowser is able to begin the process of divination and discovery as and when they please, without the necessity for a great deal of preparatory work. As such the practice is readily at hand, providing an almost instant engagement with the landscape and rapid method for archaeological discovery that swiftly brings about material results. Williamson and Bellamy claimed that a growing acceptance of leylines came as a direct result of the apparent readiness in which ley hunting offers itself as an evidence based practice: "A ley can easily be found and confirmed in a weekend so the ley hunters can feel themselves to be pushing forward the boundaries of knowledge without giving up an enormous amount of time" ([46], p. 202). The readiness of dowsing, its capacity to become a tool through which we can access and map the environment, both in terms of its past and its sacrality, solidifies its position as a methodology for getting into place. The strange way in which ordinary materials (copper rods, hazel, willow) gain agency through such a practice as this, displays a movement towards enchantment of both tools and landscape; where the ordinary becomes extraordinary.

Such mytho-geographic visions of landscape can be acted on: both Watkinsian and post-Watkinsian readings of leylines call for a getting back into place. Leylines are by their very nature experiential, one can hunt for archaeological evidence or a spiritual nurturing in the same place and both are achievable within a marginally small time frame. Each leyline experience is individualised, each track distinct and varied [23]. As such, leylines allow for an idiosyncratic connection to and accessibility of place, an acutely personal mapping of the environment.

4. Walking/Mapping/Seeing

For Watkins there was no method of identification for leys that could not be practiced by a layman once they were out in the field. In fact Watkins had given instructions in The Ley Hunters Manual in the penultimate and aptly named chapter, "Working Instructions" ([26], pp. 86–97). In this latter part of the text, Watkins carefully sets out the didactics of ley hunting, based upon a logical system of identifying sighting points—points of archaeological significance that correlate to a linear formation—and the mark points that lay in between. Mark points, Watkins describes as including mounds, moats, mark-stones, castle keeps on mounds, and beacons [26]. Following the identification of mark points, Watkins then implores the novice ley hunter to look for further evidence of a leyline in the form of confirmation points; churches, crosses, fords, tree rings, copses, single trees, camps, cross roads, and ponds [26].

In his early writings, Watkins had detailed the method by which one went about locating and tracking a leyline, he writes:

> Taking all the earthworks mentioned, add to them the ancient churches, all moats and ponds, all castles (even castle farms), all wayside crosses, all cross roads or junctions which bear a place name, all traditional trees (such as gospel oaks), marked on maps and legendry wells. Make a small ring around each on a map. Stick a steel pin on the site of an undoubted sighting point, place a straight edge against it, and move it round until several (not less than four) of the objects named and marked come exactly into line. You will then find on that line fragments here and there of ancient roads and footpaths, also bits of modern roads confirming it. Extend the line into adjoining maps, at it will usually terminate at both ends in a natural hill of mountain peak, or sometimes (in the later examples) in a legendary well or other object ([23], p. 11).

The rules set out here are extensive and define the precise manner in which one should set about locating a leyline. There are further instructions on identification in Watkins' guide, such as photographic illustration and the researching of place names. Watkins' most important instruction can be found under a subsection headed "Fieldwork", where Watkins writes:

> Experience and practice brings an insight, which quickly spots a ley. Often one can be first seen on the map, but I more often see it out of doors in "the ley of the land" itself, and this before the mark points are found ([26], p. 94).

With this in mind I made the decision to identify and "experience" a ley for myself, to attempt a "deep mapping" of sorts, providing a closer engagement with Watkins' work and the networked landscape he describes. Making use of an Ordnance Survey (OS) map and an alignment guide [47] to initially select a ley, I arrived upon a short line reaching between two points of historical importance in north Norfolk, UK: Cawston Duel Stone and Thursford Church. The map showed that, for the most part, a traceable line ran through woodland and fields, providing unrestricted access to the linear pathway. Not only was the line clear for navigation but it contained a series of sighting points such as chapels, ruins and monuments; part of my mapping experience of the leyline would be to look for further confirmation points. In doing so I could work between the OS map and the landscape through which to draw my own conclusions as to the alignment's validity and the experience(s) it would afford me.

The leyline that I selected and intended to analyse ran southeast from the village of Thursford to the site of the Duel Stone, close to nearby Cawston, another small village in the same county. The route was approximately thirteen miles in distance and between a four and six hour walk without pause. The Duel Stone itself is a heritage site that, according to the plaque located at the site, commemorates the death of Sir Henry Hobart following a duel at Cawston Heath in 1698. I selected this specific leyline for reasons of distance and accessibility; its short distance made for a manageable journey that could be thoroughly documented. Furthermore the ley ran almost entirely uninterrupted and so it was possible to remain faithful to the "old straight track", giving a detailed insight into the historical features of the leyline and further describing its affective properties—the ways in which the line would stir emotion, manipulate vistas and force composures of the body. To my knowledge, there is no existing literature on this ley, unlike others that have been identified in the Norfolk area [47,48], therefore the likelihood of repetition of data and the possibility of auto-suggestion could be minimised.

Having decided upon the leyline I would walk, I took a map, compass, ruler and a copy of Watkins' Leyhunters Manual [26] and identified the relatively short section of leyline running between the site of the Duel Stone, Cawstone and St. Andrew's Church, Thursford. The church was chosen as the end point as no official site had been recorded (or even) suggested in the existing literature on leylines in this area of the UK. I decided upon the church as being a probable Watkinsian confirmation point along the ley's course. Mapping the trajectory of the ley on the OS map, I noticed that it ran through a significant number of villages, forests, copses and Mare's-nests (small groups of trees planted for commemorative purposes). On successfully locating the Duel Stone, I found that it was nestled within such a copse. The surrounding trees proved problematic in locating my start point as the view of the Duel Stone was obscured from the roadside. A short interrogation of the local petrol station cashier confirmed that I was on the right track; the stone lay just a few meters along from the Woodrow Garage on the Norwich Road.

Arriving at the site of the Duel Stone (see Figure 1), I had no sense of having found anything of great significance, but there was nevertheless some relief in having found the site before midday and a sense of anticipation, as this was the point from which my journey would begin. It was slightly disappointing, though, that nothing struck me as unusual about the place—a place which I had genuinely hoped would be replete with some sort of unfathomable energies. A tortured groan emanated from the hinges of the neglected wooden gate that acted as the barrier between the solitude of the monument and the movement of the city bound road. Walking across unkempt grass, I made my way towards the Duel Stone where I planned to take my bearings, in order to determine the direction of the leyline and therefore my route of passage. However, the compass appeared unable to find north, providing me with different readings at each corner of the monument. Standing away from the leyline's start point I eventually managed to find my path, heading out through the right hand corner of the monument's enclosure and into the countryside along a well-forged footpath. The grasses, nettles and weeds had been beaten well into submission along what appeared to be a frequently used course of movement. I had begun to map "deeply", navigating a route by studying, moving, and sensing my surroundings, by noticing the way a path "unfolded" before me.

This initial encounter with the stone, as something capable of influencing my experience of the site, had proved a little unnerving. I later decided that it was most likely the magnetism of the stone that had interfered with the compass reading, but even if this was the case, could there be a better beginning to a journey that was endeavouring to embed the traveller within a landscape of uncanny material agency?

The Duel Stone itself, a grey tombstone-like monument, was enclosed behind iron railings and was that day set within a backdrop of clouded skies and endless fields, straddled by trees at either side. It occurred to me that the stone, having impeded upon the action of the compass, might be far more strange than I had first thought; there was the small matter of the monument being guarded not just by the railing, but also by the trees and the perimeter fencing. It occurred to me that a lot of trouble had been spent guarding something that people would seldom come across, let alone wish to damage in any way. I let my mind wander to the reasons for taking such protective measures as I headed toward the back of the enclosure and out onto the footpath leading across the field.

Figure 1. The Duel Stone, a small stone sculpture commemorating the death of Sir Henry Hobart following a duel at nearby Cawston Heath in 1698 (Image source: taken by author).

Following the path in a northwest direction, away from Cawston and the Duel Stone, I headed through the tree-lined fields where my path was already laid before me. I took this pre-existence of the route as another one of Watkins' confirmation points; "When at least half-a-mile of road or track of ancient name aligns on a ley, it has value" ([26], p. 90). This particular section of path corresponded with the area where the ley was supposed to be and, having led directly from a significantly old monument, I took this as evidence of something more "substantial" than what had initially appeared to be just a pathway. As Watkins' instructions had suggested, the path did take me across a number of small ponds, long straight roads and old, if not ancient, trees. Of greater importance was the manner in which the track composed my views: in fact at the point of reaching Salle, the second village along the ley from Cawston, the once surrounding woodland gave way to open fields and extensive vistas across the countryside. This provided me with enough of a view to see at least as far as the next three villages along the leyline and moreover, their churches, which appeared to have been built in alignment with that of my previous destination and the Duel Stone before it. If one was to follow the theoretical trajectory of settlements or sacred sites being built to a specific geometric pattern, then this had offered some affirmation. Furthermore, it brought to mind the idea that these sacred sites were connected, navigable and sited as positions to walk between as a network rather than independently.

The sightlines gained from Salle did as much to reaffirm this sense of connection as Salle's church itself. Seemingly disproportionate in scale for such a modestly sized parish, The Church of St. Peter and St. Paul loomed over the deserted village green, a structure of gigantic proportions. The building appeared out of place for a settlement so small, with a population of noticeably few residents. Furthermore, a beacon stood in the church's shadow. For Watkins, beacons were yet another indicator of an ancient track way. The ley had begun to pick up strength in its ability to affect my perception of the landscape; I was being coerced into the networked lineation of churches and trees in order to see their connection, not just to each other but also to the straight pathway that lay before me; a complete experience was emerging from the landscape. The walking of the line was allowing me to notice, to observe (and record) the "grain" of the landscape [1]. The meshing of history, sacrality and nature into a single straight line was overwhelming. The ley was beginning to make "sense" and I felt that like Watkins, I too was starting to see the relationality between place and to see landscape with new eyes.

Continuing along the route towards Wood Dalling, I passed a ringed moat to the side of a field; yet another possible Watkinsian confirmation marker. The next field I came to was being sprayed with pesticide, the land to each side was fenced off. I was forced to head back and join the road, which I followed until the next destination, Craymere Beck via Hindolveston. At times the road itself began to take the form of a straight track, reaching almost a mile of undisturbed length in either direction at points (see Figure 2). Again, this worked to confirm the path of a Watkinsian ley; Watkins considered straight roads to be an evolution of an original alignment, a continuation of the ancient track way in the modern world [40].

Figure 2. Watkins claimed that straight roads were an indicator of the evolution of an existing leyline (Image source: taken by author).

Again, the straight road that led to Craymere Beck was significant not simply because it followed the ley in a direct and more or less faithful manner, but because of the impact it had upon me as a walker. The tree-lined road produced a tunnelled vision of the landscape before me. Having no option but to walk straight along the path ahead, my surroundings forced vistas upon me; the unfolding environment appeared to have brought me its own horizon. Being coerced to follow the pathway ahead only amplified the earlier feelings that the materialities around me were being set to work somehow. Like the Duel Stone, the place itself was inflecting my vision of the landscape, or at least manipulating it. I could only follow the road in one direction or the other and so my perception of the environment was being shaped for me.

Having been forced onto the road by the spraying of the fields took me to a further point along the leyline than I had thought, although missing out on the experience of having remained faithful to the route in its entirety, the trajectory of the tarmac I ended up on only re-emphasised the encounter as one of a deep connection with my surroundings. The seemingly mundane had begun to speak to me; trees, road names, pathways and churches all provided an alternative understanding of site and of mobility, working together to form a biography of place—a narrative, both real and imagined. The leyline, or rather my embodiment of it, had become "emplaced"; the site(s) became a point of "human experience, constructed in movement, memory, encounter and association" ([49], p. 15). The perambulation of this route had seen me become thoroughly immersed in and along the ley-place.

The walk continued into the late afternoon. The line passed through another two villages before completing its course at Thursford. Both of these villages contained churches that appeared to be in correlation with the leyline. On completion, the walk seemed to have passed by with an uncanny speed, notwithstanding the impact that the first-hand experience of how sites could be connected to each other through Watkins' alignments had made upon my understanding of placial relationality.

The leyline may not have been enchanted in any truly mystical sense of the term but there had certainly been moments, if not an overall sense, of something beyond the normative, a deeper connection with place. The demarcations of normalcy and the vernacular had been shifted, what had essentially been a thoroughly researched and lengthy ramble through the Norfolk countryside had resulted in new kind of appreciation for geomantic readings of place and space. Tracking the leyline had allowed me to enter the landscape as a place "sedimented in (both) history and sentiment" ([49], p. 38). There remains a sense that the walk that day has opened up a different way of knowing place and moreover, the connections between places.

Furthermore, the way that the path had produced predisposed vistas and seemingly pre-empted the composing of the body, suggested something more than Watkins' evidence of a prehistoric track way; it pointed to the potential for a far more embodied experience of the landscape. This was a type of embodiment that could be explored further through those more esoteric readings of leys seen in the works of Michell [30] and Cope [42] In such a light, it stood to reason that there were perhaps other ways of exploring the leyline in my research, means by which practices other than walking alone might be employed to access the vital materiality of place and which might be of a far more extraordinary trajectory.

5. Conclusions

In summary, walking a leyline allows for an intense sense of mobility; an act that encourages a moving in and between places. It is a pilgrimage of sorts: a way of revitalising one's connection with the environment. Ley hunting creates a "deep map" of the world around us—a map with paths that revolve around embodiment, allowing us to explore the contours of narrative and place. It is a practice that offers us an opportunity to enrich not only our knowledge, historical and topographical, but also our awareness of place through the promoting of an acute connection with our surroundings.

Ley hunting puts place (and places) at the centre of our worldly encounters, of spiritual engagements with our surroundings, and gives rise to opportunities to connect with and perform in the landscape. Through Watkinsian traditions—the hunting, tracking, walking of leys—the landscape becomes permeated with the sensing of the past. Through practicing Watkins' methods, the environment becomes part of a vital and tangible history; the landscape comes to provide "an ancestral map for human activity" ([49], p. 38).

Whether or not leylines exist, or whether there is any scientifically justifiable evidence for believing in them is of little importance to a practice that can be viewed as being far more about land-based opportunities than anything else; opportunities to re-place the body in nature, to journey between sites of historical importance using an innovative guidance technique and moreover, the opportunity to question the solidity of our own perception. As Charlesworth states: "leys exist alright: in the imagination, on maps, sometimes on the ground" ([50], p. 140). The concept of ley hunting is one of uncovering, of digging up the not-quite-hidden past. As such it becomes something of a post-processual archaeological method, designating "a set of approaches to the ruined material past which foreground interpretation, the ongoing process of making sense of what never was firm or certain" ([1], p. xvii).

As an archaeology of sorts, ley hunting can get us back into place through a positioning of the past as a discernible feature of the present; material remains found in the forgotten tracks, monuments, trees, moats and various other sight points that Watkins instructs us to observe. The trajectory of the ley is, as we have seen, suitable for the appropriation of ancient histories. Robert MacFarlane comments on the concept of such an empty pathway as "promising events over the horizon" ([51], p. 295), one might argue that the leyline, then, offers us a vista beyond that point.

Acknowledgements

I would like to thank Sheila Hones (University of Tokyo) for her valued input, proofreading and support in the writing of this paper.

Conflicts of Interest

The author declares no conflict of interest.

References and Notes

1. Mike Pearson, and Michael Shanks. *Theatre/Archaeology*. London: Routledge, 2001.
2. Nadia Bartolini, Robert Chris, Sara MacKian, and Steve Pile. "Psychics, crystals, candles and cauldrons: Alternative spiritualities and the question of their esoteric economies." *Social & Cultural Geography* 14 (2013): 367–88.
3. Deborah Dixon. "A benevolent and skeptical inquiry: Exploring 'Fortean Geographies' with the Mothman." *Cultural Geographies* 14 (2007): 189–210.
4. Julian Holloway. "Institutional Geographies of the New Age movement." *Geoforum* 31 (2000): 553–65.

5. Julian Holloway. "Spiritual Embodiment and Sacred Rural Landscapes." In *Country Visions*. Edited by Paul Cloke. Harlow: Pearson Education Limited, 2003, pp. 158–75.
6. Julian Holloway. "Make-Believe: Spiritual Practice, Embodiment and Sacred Space." *Environment and Planning A* 35 (2003): 1961–74.
7. Julian Holloway. "Enchanted Spaces: The Séance, Affect, and Geographies of Religion." *Annals of the Association of American Geographers* 96 (2006): 182–87.
8. Julian Holloway. "Legend-tripping in spooky spaces: Ghost tourism and infrastructures of enchantment." *Environment and Planning D: Society and Space* 28 (2010): 618–37.
9. Julian Holloway, and James Kneale. "Locating haunting: A ghost hunter's guide." *Cultural Geographies* 15 (2008): 297–312.
10. Sara MacKian. *Everyday Spirituality: Social and Spatial Worlds of Enchantment*. Basingstoke: Palgrave Macmillan, 2012.
11. David Matless. "Nature the Modern and the Mystic: Tales from Early Twentieth Century Geography." *Transactions of the Institute of British Geographers* 16 (1991): 272–86.
12. David Matless. "A geography of ghosts: The spectral landscapes of Mary Butts." *Cultural Geographies* 15 (2008): 335–57.
13. Steve Pile. *Real Cities*. London: Sage Publications Ltd, 2005.
14. Steve Pile. "Spectral Cities: Where the Repressed Returns and Other Short Stories." In *Habitus: A Sense of Place*. Edited by Jean Hillier and Emma Rooksby. Aldershot: Ashgate Publishing Group, 2005, pp. 235–57.
15. Steve Pile. "Distant Feelings: Telepathy and the problem of affect transfer over distance." *Transactions of the Institute of British Geographers* 37 (2012): 44–59.
16. James Thurgill. "Immaterialising Things." *Paperweight: A Newspaper of Visual and Material Culture* 1 (2011): 6.
17. James Thurgill. "Enchanted Geographies: Experiences of place in contemporary British Landscape Mysticism." Unpublished manuscript, last modified November 2014. Microsoft Word File.
18. James Thurgill. "Mobilities of Magick." In *Sacred Mobilities: Journeys of Belief and Belonging*. Edited by Terry Gale, Avril Maddrell and Alan Terry. London: Ashgate Press, 2015, pp. 53–70.
19. John Wylie. "The Spectral Geographies of W.G. Sebald." *Cultural Geographies* 14 (2008): 171–88.
20. David Matless. *Landscape and Englishness*. London: Reaktion Books Ltd, 1998.
21. Alfred Watkins. *The Old Straight Track*. London: Methuen, 1925.
22. Hillaire Belloc. *The Old Road*, 2nd ed. London: Constable and Company Ltd, 1911.
23. Alfred Watkins. *Early British Trackways, Moats, Mounds, Camps and Sites*. London: Simpkin, Marshall, Hamilton, Kent and Co, 1922.
24. Robert Shoesmith. *Alfred Watkins: A Herefordshire Man*. Hereford: Logaston Press, 1990.
25. Adam Stout. *Creating Prehistory: Druids, Ley Hunters and Archaeologists in Pre-War Britain*. Malden, Massachusetts and Oxford: Blackwell Publishing, 2009.
26. Alfred Watkins. *The Ley Hunter's Manual*. Wellingborough: Aquarian Press, 1977.

27. Paul Devereaux. *Earth Memory*. London: Quantum, 1991.
28. Tim Ingold. *A Brief History of Lines*. London: Routledge, 2007.
29. Steven Daniels. "Lines of Sight: Alfred Watkins, Photography and Topography in Early Twentieth-Century Britain." *Tate Papers*, 1 October 2006. Available online: http://www.tate.org.uk/research/publications/tate-papers/lines-sight-alfred-watkins-photography-and-topography-early (accessed on 8 August 2011).
30. John Michell. *The New View over Atlantis*. London: Thames and Hudson, 1989.
31. Guy Underwood. *The Pattern of the Past*. London: Museum Press, 1968.
32. Adrian J. Ivakhiv. *Claiming Sacred Ground: Pilgrims and Politics at Glastonbury and Sedona*. Bloomington and Indianapolis: Indiana University Press, 2001.
33. Graham Harvey. *Listening People Speaking Earth: Contemporary Paganism*, 2nd ed. London: Hurst and Company, 2007.
34. Joanne Parker, ed. *Written on Stone: The Cultural Reception of British Prehistoric Monuments*. Newcastle-upon-Thyne: Cambridge Scholars Publishing, 2009, pp. 1–9.
35. John Foster-Forbes, and Iris Campbell. *Giants, Myths and Megaliths*. Exeter and London: William Chudley and Son, 1973.
36. T. C. Lethbridge. *Gogmagog: The Buried Gods*. London: Routledge and Keegan Paul, 1957.
37. T. C. Lethbridge. *Ghost and Diving Rod*. London: Routledge and Keegan Paul, 1963.
38. Nigel Pennick. *The Ancient Art of Geomancy: Man in Harmony with the Earth*. London: Thames and Hudson, 1979.
39. Nigel Pennick, and Paul Devereux. *Lines on the Landscape*. London: Robert Hale, 1989.
40. Paul Devereux. *The New Ley Hunter's Guide*. Glastonbury: Gothic Image Publications, 1994.
41. Paul Devereux. *Sacred Geography: Deciphering Hidden Codes in the Landscape*. London: Octopus Publishing Group, 2010.
42. Julian Cope. *The Modern Antiquarian: A Pre-Millennial Odyssey through Megalithic Britain*. London: Thorsons, 1998.
43. Danny Sullivan. *Ley Lines: A Comprehensive Guide to Alignments*. London: Judy Piatkus, 1999.
44. British Society of Dowsers. "About the British Society of Dowsers." 6 January 2013. Available online: https://www.britishdowsers.org/about (accessed on 2 February 2014).
45. Colin Wilson. *Mysteries: An Investigation into the Occult, the Paranormal and the Supernatural*, 3rd ed. London: Panther Books, 1985.
46. Tom Williamson, and Liz Bellamy. *Ley Lines in Question*. Tadworth: World's Work Ltd, 1983.
47. Shirley Toulson. *East Anglia: Walking the Ley Lines and Ancient Tracks*. London: Wildwood House Limited, 1979.
48. John Timpson. *Timpson's Leylines: A Layman Tracking the Leys*. London: Cassell and Co, 2000.
49. Christopher Tilley. *A Phenomenology of Landscape: Places, Paths and Monuments*. Oxford: Berg Publishers, 1994.
50. Michael Charlesworth. "Photography, the Index, and the Nonexistent: Alfred Watkins' Discovery (or Invention) of the Notorious Ley-lines of British Archaeology." *Visual Resources* 26 (2010): 131–45.
51. Robert MacFarlane. *The Old Ways*. London: Penguin Books Ltd, 2012.

Mapping out *Patience*: Cartography, Cinema and W.G. Sebald

Taien Ng-Chan

Abstract: Cinematic cartography can be an especially powerful tool for deep mapping, as it can convey the narratives, emotions, memories and histories, as well as the locations and geography that are associated with a place. This is evident in the documentary film *Patience (After Sebald)* by Grant Gee, which follows in the footsteps of W.G. Sebald and his walking tour of Suffolk, England, as described in his book *The Rings of Saturn*. A variety of strategies in cinematic cartography are used quite consciously in Gee's exploration of space, place and story. Using Teresa Castro's three cartographic shapes of cinema, I structure an analysis of the film's opening scene through a discussion of cinematic cartography, or the plotting of geospatial data onto a map, as well as what I will differentiate as cartographic cinema, or the mapping of space through the cinematographic image. I argue that both are necessary not only to have a deep understanding of the world and our place in it, but also in how to transmit that knowledge to others.

Reprinted from *Humanities*. Cite as: Ng-Chan, T. Mapping out *Patience*: Cartography, Cinema and W.G. Sebald. *Humanities* **2015**, *4*, 554–568.

1. Introduction

Maps help us to locate ourselves in time and space, but concepts about what a map is have changed over the centuries. Michel de Certeau describes the function of the map as moving over time from the marking out of itineraries ("performative indications chiefly concerning pilgrimages") to the colonization of space ([1], p. 120). He suggests that the map has now become an authority on "place" where there are sets of rules and plans, streets and architecture, points of interest, whereas "space", which is the tour, the narrative, the context, and the human perspective, has disappeared from the map. Similarly, Robert Macfarlane describes what he terms "story maps" and "grid maps" as the difference between imaginative and functionalistic. Macfarlane defines story maps as "forms of spatial expressions that embody our personal experiences of the environment and contribute to creating a deep understanding of places" ([2], p. 142), whereas grid maps train the imagination to see and think in so-called "scientific" ways that purport to be "objective." However, Doreen Massey critiques the construction of binaries such as "space" and "place" (or "story" and "map"), suggesting instead that space be seen as "the product of interrelations…always under construction" and as "a simultaneity of stories-so-far" ([3], p. 9).

Strategies of "deep mapping" or "deep topography" hold much potential to expand spatial knowledge to inter-related and simultaneous connections of "stories-so-far", particularly if extended to cinema. These terms usually refer to practices of landscape-based writing that can incorporate history, (auto)biography, folklore, memory, and all types of narratives and emotions, and that may manifest as any combination of literature and illustration, as well as radio, performance, and multimedia [4–7]. Denis Wood points out, however, that these "deep maps" are more metaphorical than cartographic; he also calls into question the designation of "deep" mapping *versus* "thin" mapping [8]. As Wood has shown, all maps tell stories [9], and I do not wish to set up an opposition but rather a description (just as deep and shallow waters are not opposed but descriptive of depth) of how a "deep map" can move beyond the "surface" imagery that a grid map provides. As Deleuze and Guattari say, "Make a map, not a tracing" ([10], p. 12)!

A recent film *Patience (After Sebald)* by Grant Gee [11] can be considered a deep map, one that demonstrates a necessary synthesis of story map and grid map, or the narrative and the functionalistic, by using both as frameworks for a cinematic exploration. *Patience* literally follows in the footsteps of writer W.G. Sebald, in modes of film essay/biography/travel documentary, by traversing the same path described in Sebald's literary work *The Rings of Saturn* [12]. It is not a direct translation of the book, but rather the director's experience of those same landscapes, mixed in with a more traditional "talking heads" documentary approach with interviews of people who knew or have been influenced by Sebald. It is the story about pilgrimage, which, as Macfarlane points out in an interview within the film, is the subtitle of the English version of the book (originally written in German). Lise Patt, editor of the anthology *Searching for Sebald* and another interviewee, makes note of the "image" in the word "pilgrimage." She points to medieval itinerary maps with images of the pilgrimages for those who did not actually go on the journey. People would use the maps to meditate upon, to move their minds through the different parts of the journey on the way to Jerusalem. Thus, in addition to being a meta-travelogue—that is, a travel story about a travel story—*Patience* functions as a kind of image-map, addressing the question of how both locational knowledge and emotional experience of place can be transmitted through cinema.

Cinematic narratives and maps have been linked since the earliest days of silent filmmaking, when newsreels and travelogues provided glimpses of distant lands [13–15]. Maps and travel films both provide a sense of possibility and accessibility to "foreign" lands that, as Tom Gunning points out, "forms a cornerstone of the modern worldview in which technology can render every distant thing somehow available to us" ([14], pp. 27–28). In addition, filmmakers and cartographers are closely related through their attempts in visualizing the world, as Teresa Castro shows in her essay "Cinema's Mapping Impulse" [16]. She identifies three "cartographic shapes" that are present in cinema: the aerial view, the panorama, and the atlas. Each of these shapes are very different: "if panoramas maximize the notion of *point view*, atlases relate to a way of *assembling* images, while aerial views concern a particular *angle of view*" ([16], p. 11). Through Castro's cartographic shapes, I propose to trace some of the functions and tropes of cinematic cartography and its relation to cartographic cinema, as they are used in *Patience (After Sebald)*. I suggest that cartographic cinema can be especially effective in relating the tour, that is, narrative, the human perspective and the movement through place, while the map remains necessary to delineate location. *Patience (After Sebald)* offers a variety of cinematic and cartographic strategies, including both the map and the tour, which, I argue, are necessary not only to have a deep understanding of the world and our place in it, but also in how to transmit that knowledge to others.

2. Cinematic Cartography/Cartographic Cinema

W.G. Sebald was a German writer who, at the time of his early death at the age of 57, had been cited by literary critics such as Susan Sontag as one of the world's greatest authors [17]. His works are a strange and unique blend of memoir, travelogue, history, and fiction, particularly concerning the trauma of the Second World War. British filmmaker Grant Gee, on the tenth anniversary of Sebald's death, released the documentary film *Patience (After Sebald)*, a cinematic pilgrimage that follows Sebald's tour of the areas around the University of East Anglia, England, where Sebald taught for much of his later life. Elements in *Patience* take off from cinematic cartography, showing the locations of Sebald's walking tour of Suffolk as well as the locations mentioned in the spiralling and divergent narrative, as plotted out on Google Maps. This use of the grid map allows for a geographical overview of place, a precise pinpointing of location. Cartographic cinema is also a device used in the film to describe architecture and landscapes, and allows the more subjective experience of place to filter through the imagery. Both of these modes of cinematic cartography are necessary to the narrative, and this documentary is perhaps the first to relate the different mapping practices together quite consciously. In fact, Gee has studied geography and even completed a year towards a Ph.D. in the field [18]. He has undoubtedly read—and even interviews in the film—some of the same writers that I draw upon in this analysis (Robert Macfarlane, for instance).

Cinematic cartography is a rapidly growing subfield of interdisciplinary study, as evidenced by a plethora of books recently published on the subject (for example, see [19–21]). The term has been used to describe many different relationships between mapping and cinema. Les Roberts, in his essay "Cinematic Cartography: Projecting Place Through Film" [22], identifies five different main approaches, or clusters, as he calls them, which are by nature inexact and overlapping. Of the five clusters mentioned, four can be found in *Patience,* the exception being the mapping of film production and consumption. Gee brings in maps and mapping in film, that is, the plotting of geospatial data onto a map, as already discussed, as well as movie mapping and cinematographic tourism; cognitive and emotional mapping; and film as spatial critique. In this essay I suggest a differentiation between cinematic cartography as specifically the use of maps to plot out a film's spatial data, and cartographic cinema as the exploration of space through the moving image, which often includes cognitive and emotional mapping and film as spatial critique.

Castro addresses cartographic cinema, particularly how both cartography and cinema share "a particular way of seeing and looking at the world, a visual regime" ([16], p. 10). She draws on Svetlana Alper's book *The Art of Describing* [23], which shows that seventeenth century visual culture in the Netherlands, including maps and paintings, were historical modes of thought impacting how space is represented in modernity. Through an investigation of three cartographic shapes, Castro also links these views to war, as they were made possible by wartime technology and used for military applications. This has relevance to today's digital maps that are made possible through satellite technologies, also invented for military purposes. The themes of landscape, history, and war are all prevalent in Sebald's work, and further support the usefulness of Castro's three cartographic shapes to this analysis. I also extrapolate their significance to modern digital technologies, including online mapping services such as Google Maps and Google Street View.

And finally, I suggest that the combination of both grid and story map can be achieved through cinema as an extremely useful tool for the practice of deep mapping, particularly as a compelling way to associate emotions, stories and histories with geographic locations.

3. Cinematic Cartography and the Mapping of Location

Maps have often appeared in films to provide locational grounding and to lend realism or metaphor to the narrative, whether documentary or fiction [13]. However, it is only more recently that cartography has included the practice of plotting a film's—or any other kind of narrative's—locations on a map. As Sébastien Caquard in "Mapping Narrative Cartography" notes, "maps are more and more commonly used as an analytical tool to explore the spatial dimension of narratives, marking the emergence of hybridized practices" including literary and cinematic mapping ([24], p. 135). Barbara Piatti *et al.* point out that there are examples of literary maps dating back to the beginning of the 20th century, for instance, William Sharp's map of Scotland depicting the zone of action for several novels by Walter Scott. However, they argue that these early experiments in literary geography were rudimentary and illustrative of other written texts, without having much weight of their own, and maintain that it has only been since Franco Moretti's work *Atlas of the European Novel* (1998) have narrative maps "become truly tools of interpretation, allowing to see something which hasn't been evident before" ([25], p. 180).

In *Patience (After Sebald)* the very first image that we see is a Google Map with placemarkers and lines stretching out all over the world (see Figure 1).

We begin to zoom in to England, where the majority of placemarkers are found, to the sound of birds and the crack of a rifle. We keep zooming in, jump cuts simulating the redrawing of the computer screen that happens in Google Maps. We begin to make out details in the satellite images of the land. As buildings and roads take shape, a placemarker pops up, reading "Norwich", where Sebald's book *The Rings of Saturn* begins (see Figure 2).

Figure 1. Screenshot from the opening sequence of the film *Patience (After Sebald)* by director Grant Gee (2011), showing placemarkers and connecting lines on the world map.

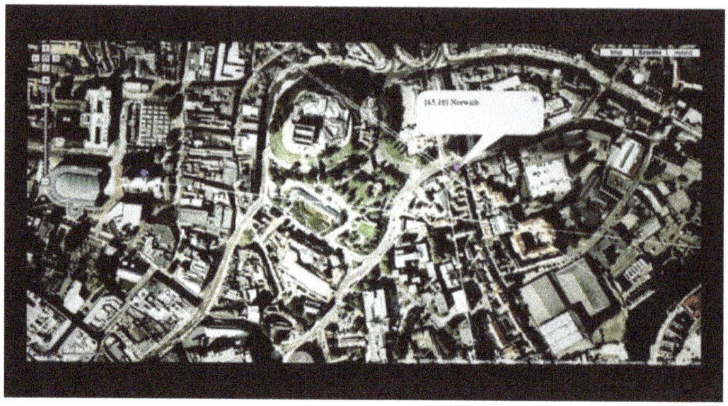

Figure 2. Screenshot from the opening sequence of the film *Patience (After Sebald)* by director Grant Gee (2011), showing "Norwich".

Barbara Hui, writer and creator of the Litmap project, is one of the interviewees in *Patience*, and it is her Google Maps mashup that opens and frames the film's narrative (a mashup combines data or content from several different sources). Hui relates how, since Sebald's book is full of geographic place names, she decided to plot each place using the Google Maps Application Programming Interface (API) tool. Hui uses the colour red on the map to indicate places from Sebald's actual walking tour, with links out to the places that he mentions as his thoughts are roaming. This use of maps in the film works doubly as literary cartography that plots the locations of a work of literature and cinematic cartography about where the film was shot as well.

This instance of narrative cartography conveys an immediate and explicit spatial dimension to the analysis. One is able to see the lay of the land, the distances involved, how things are in relation to each other, the shape of the coastlines, the topography, the outline of the journey. These are all extremely helpful things in setting a film or fleshing out any kind of narrative, and maps often appear in films for these reasons. The use of the digital map in *Patience* also indicates how many place names are mentioned in the book's narrative, and that these are very specific locations of the world, aiding in the documentary aspects of the film. And finally, as Hui points out, "in *The Rings of Saturn*, Sebald illustrates a spatialized view of history that sees the local as globally defined" [26]. This is shown graphically through the use of connecting lines that join distant countries to the places of Sebald and Gee's walking tour. Local history cannot be disassociated from global history.

Hui acknowledges, however, that there are certain challenges and limitations with her project, and with narrative cartography in general. For instance, she points out that "Sebald has a 'cosmological' notion of historical space in addition to the local and global one. It is not possible to map this cosmological notion of space using the Google Maps API!" [26]. Sebald's writing is a hybrid of travelogue, autobiography, history, and fiction, among other genres, and as Piatti *et al.* have pointed out, there are particular challenges in mapping fictions, even those based in real world counterparts [25]. Locations may be vague, partly or completely invented. In addition, Hui points out that the use of Google Maps or similar "restricts the user to that vision/version of geographical space. When you're working with that data, be aware that you're working with those pre-provided layers of information, which have their own inherent assumptions built in" [26].

Wood calls maps "arguments about existence" ([9], p. 34). They tell stories through what is included and what is left out. Colonial maps, for instance, helped to shape nations and states by presenting them as united by imaginary borders [27]. Instead of colonial interests, the digital grid map, most popularly represented by Google Maps, now depicts commercial ones, a view of the landscape as places of consumption. The Google satellite view seen in the film thus adds other layers of meaning, with its scientific overtones and notions of indexicality. It gives as well the pleasure of being able to see the whole Earth from an exotic and in fact impossible angle. It places a grid upon the world, so that anything can be found according to exact coordinates of longitude and latitude, thus, the term "grid map" as advanced by Macfarlane.

The grid map of Google's satellite view corresponds with Castro's aerial view, in that both strive through photographic means to make accurate plans of the land from above. They both aim to make landscapes into maps as accurately as possible, and they are both associated with military uses as well as pleasurable and thrilling emotions. Castro, in her discussion of the aerial view as a cartographic shape, looks at films shot from balloons, dirigibles, and airplanes, which were part of an increasingly popular theme of travel during the early part of the 20th century. She points out that the Lumière brothers had made films shot from hot-air balloons as early as 1898. Castro focuses particularly on one film, *En dirigeable sur les champs de bataille* (1918), which was made by the Cinematographic Service of the French Army (Service Cinématographique de l'Armée). Shot from a dirigible shortly after the First World War, the camera flies over the combat zones of Northern France, exposing precise quantities and locations of the devastation, while "the smoothness and fluidity of its aerial movement represent an unquestionable source of emotion" that includes the visual pleasure of a novel angle of vision, the horror at the ruins of war, and the "e-motion" of the feeling of flying through space ([16], p. 14). Such "airplane maps" were part of a larger post-war documentary and cartographic project that needed accurate plans in order to progress in the reconstruction. In addition to an inventory of the land, such images also acted as nationalistic propaganda "in particular because of their insistence on the image of a sacrificed land and on the urgency of the rebuilding campaign" ([16], p. 14).

The aerial view pre-dates the satellite view, of course, but anticipates how science, accuracy, and representational economies (particular scopic regimes) have transformed the world into the gold of "tradeable values" ([28], p. 22). The satellite view is able to take the aerial view to the extreme, framing the entire planet in its view. There have been incredible advances and benefits made possible by such grid maps, but while they are powerful tools, as Macfarlane points out, "their virtue is also their danger: that they reduce the world only to data, that they record space independent of being" ([2], p. 141). One might see the grid map as useful for locating and positioning and for mapping surfaces, but inappropriate for mapping the deeper layers of story, emotion and history. The Google Map can tell me where Norwich, East Anglia or Somerleyton Hall is, geographically, but otherwise, to me, not being familiar with the area, they are just names, placemarkers, or points on the map.

How then to trace and convey the narratives and contexts without losing location and cartographic functionality? As Sébastien Caquard and Fraser Taylor note, the diversity and power of cinematic approaches to evoke emotional, political and personal dimensions of a place contrasts with the scarcity of those dimensions in cartographic practices [29]. A hybridized cartographic cinema can help map and convey the affective experiences of place, as the next two sections will explore.

4. Cartographic Cinema and the Mapping of Space

From the Google Maps mashup in the opening shots of *Patience*, we move to a long horizontal tracking shot of an East Anglia train, which we would infer (if we are following along with book in hand) is leaving from Norwich, bound for Somerleyton Hall, Sebald's first stop (see Figure 3). The film cuts from the satellite map of the world to the street level point of view on the train station, encapsulating the movement to the second of the three "cartographic shapes" of the film: the panorama. A beautiful black and white extended tracking shot travels the length of a train, the words "East Anglia" on the sides of the coaches, giving way to travelling shots from the train, of the tracks and the countryside.

Figure 3. Screenshot from the opening credits sequence of the film *Patience (After Sebald)* by director Grant Gee (2011), showing the panoramic viewpoint.

Tom Conley, in his book *Cartographic Cinema*, contends that "[e]ven if a film does not display a map as such, by nature it bears an implicit relation with cartography" ([30], p. 1). The panorama is evidence of this. Castro describes the panorama shape as responding to "a desire to embrace and to circumscribe space, allowing for the observer's eye to seize the whole of an image" ([16], p. 11). She refers to early cinema's penchant for capturing views and landscapes that represented "the careful scaling and coding of the world through filmic means, namely, horizontal and 360° panoramic shots" ([16], p. 12). While there are few actual panoramic shots in *Patience* (that is, where the camera turns or pans around, rotating on a tripod axis), there are numerous horizontal tracking shots that guide the spectator through time and space. Even when there is no camera movement, the landscapes that Gee films with his 16mm Bolex invite the connection to 18th and 19th century painting as well as modern landscape photography, invoking the spectacularization of landscape and the associated longing for immersion in and mastery over space. Guiliana Bruno describes, for instance, how view painting "merged the codes of landscape painting with urban topography" ([31], p. 61), which has been absorbed into the language of film. "Travel culture is written on the techniques of filmic observation" ([31], p. 62).

Since the very idea of travel is made possible through maps and the provocation of the imagination, tourism, consequently, has become about ways to consume the images of places depicted as spectacular views. As Gunning points out, travel films played a major role in the selling of early cinema with its emphasis on spectacle and attraction, but these films were only the latest in a tradition of traveling lectures with slide shows, moving landscape paintings, and the stereoscope with its illusionary depth. He argues that "[t]he idea of making the world available in the form of a spectacle stretches back to the panorama of the end of the eighteenth century and to the world expositions of the latter part of the nineteenth" ([14], p. 29). In the modern era, the concept of travel is now more than ever bound up with the production, commodification, and consumption of images, including travel shows, postcards, selfies, and maps, all of which do not simply function as substitutes for those who cannot travel, or souvenirs for those who have. They also work to structure the imagination of travel and the very experience of the journey itself, by providing a "script" of expectations.

Where the aerial view has its modern counterpart in Google Maps, satellite imagery and grid maps in general, the shape of the panorama has its modern cartographic parallel in Google Street View (see Figure 4). Google Maps launched its Street View feature in May 2007. "The underlying idea is very simple: Provide an interface that can display street-level images in a natural way that enables convenient navigation between images without losing the map context" ([32], p. 118). Street View provides 360 degree street-level panoramas of many cities, mainly in Western countries, as well as popular tourist sites around the world. And in May 2012, Google introduced the World Wonders Project, a series of Street View tours of such famous historical sites as Stonehenge, the Versailles Palace, and Sir Robert Scott's hut in the Antarctic. Underlying the conception of GSV is the obsession to photograph the world and make it available as never before, and in this, the Google Street View panorama "bubble" is directly connected to the panorama paintings and photos of the late-eighteenth century, as well as to early travel films.

Figure 4. Screenshot of the *Litmap* project's website, showing a Google Street View image of the first red placemarker (Brundall Station) on Barbara Hui's *Litmap* [33].

The landscape panorama provides a counterbalance to the overhead aerial or satellite view. It represents the territory that was abstracted by the grid map; it speaks to the desire to move through map space, to fly into it and see it up close as landmarks, streets and buildings, as the animations of Google Earth illustrate. The panorama describes the space of being, that is, a more human point of view, and in a much more precise manner than words ever can—a picture is worth a thousand words, as the old saying goes. I never really knew what a moor looked like, for instance, until I saw a photo of one, even though moors are often the settings in English novels. However, as Bruno notes, a particular sense of place is actively generated by both images and narratives, as well as emotions [31]. If the aerial view is an overhead map and the panorama describes the street level perspective, both of these remain surface mappings, representations of space that have no history or context without the third cartographic shape, that of the atlas. *Patience* is, in many ways, about the interrelations between the map and the tour, and the need for narrative and meaning that is made possible through the atlas view of montage and story, as the next section will explore.

5. The Atlas as Montage and Meaning

Immediately after the film's opening sequence that sets up the mapping discourse with the overhead view and the panorama, Gee introduces the first explicit connection between a place in the book and a place in the film. This is Somerleyton Station, where Sebald and Gee get off the train. The reference in the book is given to us on the screen, along with a small rectangle of colour (of a man walking alongside a departing train—perhaps the filmmaker himself?) superimposed upon a background of grainy black and white field (see Figure 5). Gee's various visual strategies such as this superimposition, as well as the use of montage in general to organize images thematically, can be seen to correspond to the atlas as a cartographic shape. Castro's atlas refers not only to a collection or archive of images that aim to convey geographical knowledge as well as history, but also "a means to organize visual knowledge. In other words, atlases refer as much to a strictly cartographic instrument as to a graphical means for the assemblage and combination—if not montage—of images" ([16], p. 13). As Christian Jacob points out in his book *The Sovereign Map* [34], the similarities between the author of an atlas and the editor of a film reside in their use of such elements as framing, rhythm, focus, a sense of progression, and structure.

Figure 5. Screenshot from the film *Patience (After Sebald)* by director Grant Gee (2011), showing an assemblage of two different viewpoints.

The editing of landscape scenes in *Patience* often functions as assemblages and combinations that aim to convey geographical information as well as dual points of view. For instance, the black and white landscape footage is often presented to illustrate Sebald's words, while the smaller rectangle alludes to the filmmaker's experience of the same landscape through a handheld camera point of view, often of his walking feet. Elsewhere, montage sequences of cliffs, beaches, or moors give evidence of an atlas-view towards conveying different parts of a landscape. Sebald describes his own experiences of feeling both liberated and despondent whilst moving through the landscape, and often touches off from a chance encounter, a particular place or the history of that place, to spiral out into something else, an encyclopaedia of people and things that he is interested in. Gee replicates Sebald's strategy, using a combination of the panorama and the aerial cartographic shapes, as well as Hui's Google Maps mashup, to give various layers of geographical knowledge, from topography and location to distance and point of view. For instance, he reproduces Sebald's photos from the book, superimposing over them his own footage of the same place (see Figure 6).

Figure 6. Screenshot from the film *Patience (After Sebald)* by director Grant Gee (2011), showing an image of Somerleyton Hall from Sebald's book *The Rings of Saturn* superimposed on a cinematographic image of the same view.

Alone, visual strategies, whether maps or panoramas, are not enough to give a full sense of place. They are tools for the mapping impulse to describe space, and to open the way for the tour, for narrative, (auto)biography and history, which deepen the picture. This next scene in the film constructs a narrative spiral, following that of the book's, to recount Sebald's chance meeting with a gardener at Somerleyton Hall. The gardener finds out Sebald is German, and tells him about the planes setting out from the 67 airfields in East Anglia to bomb Germany. Over this, we see what looks like archival footage of warplanes and bombs, projected on a screen. This is the aerial view in the original usage of the term, which, as Castro points out, was associated with war and military operations. In the film, we see aerial footage from the East Anglian Film Archive, shot from planes dropping bombs onto a large target painted in white lines on a green field (see Figure 7).

Figure 7. Screenshot from the film *Patience (After Sebald)* by director Grant Gee (2011), showing archival footage from the East Anglian Film Archive.

This scene does several things. It introduces the themes of war and death that are prevalent in much of Sebald's work, it visually illustrates the irradiating melancholy and its impact upon the landscape—an emotional mapping—and also tells both a history of the area and a story of an encounter with a local, a chance meeting on the pilgrimage. Whether the story is factual or fictional is irrelevant, since either way, it conveys an emotional connection. The montage works to relate the different cartographic shapes together in exploring a single place, but also emphasizes the importance of the tour in relation to the map. In this depiction of East Anglia, the Second World War can be read as part of the landscape, but requires narrative and the human dimension to be fully accessible. At the same time, without the location of the place on the map, one would not have a clear idea of East Anglia's geography and its relationship to the world. Gee's montage of images and the trope of "screen projection" to connote archival imagery play in concert with Sebald's telling of a story to make emotional and historical meaning of the place. Through this cinematic deep mapping, East Anglia now means something more to me than when it was simply a point on a map.

6. Conclusions

Visual knowledge remains on the surface, a "shallow" cartography that is a "tracing" rather than a deep map. How can maps also function to transmit deep knowledge? Spectacular and well-known destinations remain tourist clichés without contexts, local points of view, and narratives, while more mundane places remain simply abstract points on a map that hold no meaning without a deep human perspective. It is becoming more and more important to find ways of expanding the tourist/traveller perspective, to provoke viewpoints about places that go beyond the consumption of images.

Castro's cartographic shapes prove useful as a framework for exploring the relationships between cinema, cartography, narrative and the deep mapping of place. The aerial view acts as a way of ordering place. It displays knowledge from an overhead perspective and can be extended to satellite imagery. The panoramic street level view opens up a more human perspective. The cinematographic image and motion functions to map the space of being. Finally, the atlas view, through montage, functions to organize these images into thematic explorations and narratives. These are tools for the mapping impulse to describe space visually, and to open the way for the tour, for emotion, (auto)biography and history that can be revealed in the landscape through deep mapping.

Sebald's narratives are linked to the landscape as well as to an associative cataloguing of thoughts and emotions. When something like a story or an evoked feeling resonates with a person, it affects that person's point of view on the landscape. As Gee suggests, "the book starts to go way beyond its boundaries; you start trying to find the traces of it in the world" [18]. In *Patience (After Sebald)*, Gee replicates Sebald's associative strategy of spiralling narrative to give his film various layers of geographical knowledge, from topography and location to associations, emotions, and stories. The beautiful landscape shots work as a cinema of attraction, giving glimpses that make us want to see these places, and the associated stories give these places meaning. The grid map gives precise locations, and enables us to replicate the journey ourselves. This is how itineraries are formed, the frameworks with which to build pilgrimages and tours.

This essay began by discussing the idea of cinema as the "image" in pilgrimage, as a kind of itinerary map. The itinerary was one of the earliest forms of widespread spatial representations for travellers, mainly used by pilgrims to get to Jerusalem—the main destination of Christian pilgrimages. Although many pre-modern itineraries were only textual, sometimes they included graphics that transformed the itinerary into an itinerary map [35]. Most people at the time could not have undertaken these journeys because of the expense and hardship involved. Therefore, these representations functioned simultaneously as practical instructions for pilgrims, imaginative instructions for non-pilgrims, and a means of engraving Christianity into the landscape by guiding the pilgrims on a common path, which encouraged the growth of services to pilgrims along these routes. Nowadays, literary and cinematic tourism has become in many ways the equivalent forms of pilgrimage. The mapping of locations that appear in novels, films and other narratives allow for people to partake in what Thierry Joliveau calls "set-jetting" [36]. This is another aspect of cinematic cartography found in Gee's film.

Iain Sinclair is interviewed in the film saying that replicating Sebald's walking tour, as many people are now doing, is "a pilgrimage because the man is dead and he is ripe for a cult…we cannot literally follow along because he's a mythologist. Many of the things in there he's invented...But why not do that walk? It's a great walk, enjoy it." Macfarlane, also in the film, talks about the British tradition of walking as recovery from the accumulations of the city, while the American tradition of walking is more as discovery. Both Sinclair and Macfarlane are known for their interest in deep mapping or similar practices, but they point out that there is only so much that one can learn from maps, cinematic or otherwise. There is no replacement for the actual experience of walking, of being there. Gee alludes to this with a sequence in the film where Hui speaks about the amazement and thrill of being able to find and visit some of the actual places in *The Rings of Saturn*, such as the Bull Inn or the "Shop in Clarahill". She is showing where these places are on Google Maps, but then, as she attempts to zoom in ever closer, the map disappears into an error message that reads "We are sorry, but we don't have imagery at this zoom level." Superimposed behind this apology, faintly, is a cinematographic image of the place from the street level view (see Figure 8).

Figure 8. Screenshot from the film *Patience (After Sebald)* by director Grant Gee (2011), showing a Google Maps error message superimposed on cinematographic footage of a street.

Gee includes this scene as a suggestion that grid maps are limited in their viewpoints. Without the grid map, the street view has no location, and cannot be visited. However, because the overhead grid map cannot give a sense of the space of being, the street level view is able provide a more human perspective. However, the best way to learn about the world is by moving through it, although that movement may be physical or mental, and deep mapping practices suggest that it is both. Maps certainly help to get one started, to get out into the world to make one's own experiences, but for those who cannot make the journey, and for those who wish for a guide, cinema as image-map can provide both location and narrative. *Patience (After Sebald)* can be thought of as a kind of itinerary map that guides one through a personal and emotional topography, both Sebald's and Gee's. It illustrates how cinema can function as an especially powerful tool for deep mapping, as it can both record and recount the narratives, emotions, memories and histories, as well as the locations and geographies that are associated with a place, for a deeper understanding of the world in which we live.

Acknowledgments

The author would like to thank the peer reviewers for their comments, and especially Sébastien Caquard for his readings of earlier drafts as well as for his general support. Thanks also to Grant Gee and Gareth Evans for permission to reproduce screenshots from their film. A shorter version of this paper's contents was presented at the 26th International Cartographic Convention in Dresden, 2013.

Conflicts of Interest

The author declares no conflict of interest.

References

1. Michel de Certeau. *The Practice of Everyday Life*. Translated by Steven Rendall. Berkeley: University of California Press, 1984.
2. Robert Macfarlane. *The Wild Places*. London and New York: Granta Books and Penguin Books, 2007.
3. Doreen Massey. *For Space*. London: Sage Publications, 2005.
4. Iain Biggs. "'Deep Mapping': A Brief Introduction." In *Mapping Spectral Traces*. Edited by Karen Till. Blackburg: Virginia Tech College of Architecture and Urban Studies, 2010, pp. 5–8.
5. Iain Biggs. "The Spaces of 'Deep Mapping': A Partial Account." *Journal of Arts and Communities* 2 (2011): 5–25.
6. William Least Heat-Moon. *PrairyErth (A Deep Map)*. New York: Mariner Books, 1991.
7. David Cooper, and Les Roberts. "Walking, Witnessing, Mapping: An Interview with Iain Sinclair." In *Mapping Cultures: Place, Practice, Performance.* Edited by Les Roberts. Hampshire and New York: Palgrave McMillan, 2012.
8. Denis Wood. "Mapping Deeply." *Humanities* 4 (2015): 304–18.
9. Denis Wood. *Rethinking the Power of Maps*. New York: The Guilford Press, 2010.
10. Gilles Deleuze, and Felix Guattari. *A Thousand Plateaus.* Translated by Brian Massumi. Minneapolis: University of Minnesota Press, 1987.
11. *Patience (After Sebald)*. Directed by Grant Gee. London: SODA, 2012. Premiered January 2011.
12. Winfried Georg Sebald. *The Rings of Saturn.* Translated by Michael Hulse. London: The Harvill Press, 1998.
13. Sébastien Caquard. "Foreshadowing Contemporary Digital Cartography: A Historical Review of Cinematic Maps in Films." *The Cartographic Journal* 46 (2009): 46–55.
14. Tom Gunning. "The Whole World Within Reach." In *Virtual Voyages*. Edited by Jeffrey Ruoff. Durham: Duke University Press, 2006, pp. 25–41.
15. Teresa Castro. "Les Archives de la Planète—A cinematographic atlas." *Jump Cut*, 2006. Available online: http://www.ejumpcut.org/archive/jc48.2006/KahnAtlas/index.html (accessed on 7 September 2015).
16. Teresa Castro. "Cinema's Mapping Impulse: Questioning Visual Culture." *The Cartographic Journal* 46 (2009): 9–15.
17. Mark O'Connell. "Why You Should Read W. G. Sebald." *The New Yorker*, 14 December 2011. Available online: http://www.newyorker.com/books/page-turner/why-you-should-read-w-g-sebald (accessed on 29 July 2015).
18. Evan Harris. "Interview with Grant Gee." *The White Review*, April 2012. Available online: http://www.thewhitereview.org/interviews/interview-with-grant-gee/ (accessed on 29 July 2015)
19. Chris Lukinbeal, and Stefan Zimmerman, eds. *The Geography of Cinema—A Cinematic World.* Stuttgart: Franz Steiner Verlag, 2008.
20. John David Rhodes, and Elena Gorfinkel, eds. *Taking Place: Location and the Moving Image.* Minneapolis: University of Minnesota Press, 2011.

21. Les Roberts, and Julia Hallam, eds. *Locating the Moving Image: New Approaches to Film and Place*. Bloomington: Indiana University Press, 2013.
22. Les Roberts. "Cinematic Cartography: Projecting Place through Film." In *Mapping Cultures: Place, Practice, Performance.* Edited by Les Roberts. Hampshire and New York: Palgrave McMillan, 2012.
23. Svetlana Alpers. *The Art of Describing: Dutch Art in the Seventeenth Century*. Chicago: University of Chicago Press, 1983.
24. Sébastien Caquard. "Cartography I: Mapping narrative cartography." *Progress in Human Geography* 37 (2013): 135–44.
25. Barbara Piatti, Hans Rudolf Bär, Anne-Kathrin Reuschel, Lorenz Hurni, and William Cartwright. "Mapping Literature: Towards a Geography of Fiction." In *Art and Cartography*. Edited by William Cartwright, Georg Gartner and Antje Lehn. Berlin: Springer-Verlag, 2009.
26. Barbara Hui. "Litmap Presentation Notes." 2 June 2009. Available online: http://barbarahui.net/2009/06/02/litmap-presentation-notes/ (accessed on 29 July 2015).
27. Martin Brückner. *The Geographic Revolution in Early America: Maps, Literacy, and National Identity*. Chapel Hill: The University of North Carolina Press, 2006.
28. John Pickles. *A History of Spaces: Cartographic Reason, Mapping, and the Geo-coded World.* London and New York: Routledge, 2004.
29. Sébastien Caquard, and Fraser Taylor. "What is Cinematic Cartography?" *The Cartographic Journal* 46 (2009): 5–8.
30. Tom Conley. *Cartographic Cinema*. Minneapolis: University of Minnesota Press, 2007.
31. Guiliana Bruno. *Atlas of Emotion: Journeys in Art, Architecture, and Film*. New York: Verso, 2002.
32. Luc Vincent. "Taking Online Maps Down to Street Level." *Computer* 40 (2007): 118–20.
33. Barbara Hui. "Litmap." Available online: http://barbarahui.net/litmap/ (accessed on 29 July 2015).
34. Christian Jacob. *The Sovereign Map.* Translated by Tom Conley. Chicago: The University of Chicago Press, 2006.
35. Catherine Delano-Smith. "Milieus of mobility: Itineraries, route maps, and road maps." In *Cartographies of Travel and Navigation.* Edited by James R. Akerman. Chicago: University of Chicago Press, 2006.
36. Thierry Joliveau. "Connecting Real and Imaginary Places through Geospatial Technologies: Examples from Set-jetting and Art-oriented Tourism." *The Cartographic Journal* 46 (2009): 36–45.

Deep Mapping and Screen Tourism: The Oxford of Harry Potter and Inspector Morse

James Cateridge

Abstract: This article proposes that the experiences of screen tourists in Oxford help to create a theoretical "deep map" of the city which explores place through narrative. Building on the travel writing of William Least Heat-Moon and other recent work in the spatial humanities, two case studies of major screen tourism drivers are considered and analyzed. The British television drama *Inspector Morse* (1987–2000) explores the ambiguity of Oxford intellectualism through its central character. Morse's love of high culture, especially music, provides suggestive additional layers for multimedia mapping, which are realized online through user-adapted *Google Maps* and geolocated images posted on the *Flickr* service. Harry Potter fans may not be "pure" or independent screen tourists, but they provide a wealth of data on their interactions with filming locations via social media such as *Instagram*. This data provides emotional as well as factual evidence, and is accumulating into an ever richer and deeper digital map of human experience.

Reprinted from *Humanities*. Cite as: Cateridge, J. Deep Mapping and Screen Tourism: The Oxford of Harry Potter and Inspector Morse. *Humanities* **2015**, *4*, 320–333.

1. Introduction

As I sit writing this article in the hallowed halls of the Bodleian Library's Radcliffe Camera reading room in the heart of Oxford, my immediate surroundings are hushed and quiet, filled with ancient books and students blinking in the light of laptop screens. By contrast, outside the window is the hustle and bustle of a major international tourist attraction. Couples, families and organized groups wander gazing at the landmark within which I am sitting, most of them making their own images of the building with cameras or mobile phones. Occasionally a cluster of sun hats pops up apparently atop one of the high walls which surround Radcliffe Square and partition Oxford's famous quads. The hats belong to tourists being guided around Exeter College, which (a quick *Google* search reveals) is the alma mater of pre-Raphaelite artist William Morris, but also the home of the perfect green lawns upon which bucolic television detective Inspector Morse (John Thaw) collapsed of a fatal heart attack. More recently, director Peter Jackson visited the college to discuss his blockbuster adaptations of J. R. R. Tolkien's *The Hobbit* (1937), written (according to *Google Maps*) 27 minutes' walk away in North Oxford. Just over the road larger groups of school children and foreign language students queue to walk up the beautiful vaulted staircase of The Divinity School in order to take photos of themselves apparently in Hogwarts. These photos will be posted immediately (and often publicly) on image sharing services such as *Flickr* or *Instagram* along with discussions about how much visiting Oxford makes you feel like you're in a *Harry Potter* film.

It is therefore possible to imagine Radcliffe Square as the symbolic centre of an imaginary map of "Oxford" which is made and remade countless times by its many visitors. Each of these visitors starts the day with their own expectations and itineraries, primed by a combination of physical guide books, prior cultural experiences—such as watching *Inspector Morse* on television—and online recommendations from sites such as *TripAdvisor*. At the end of the day they will have added a further layer of lived experience to this imaginary map, yet often even these "real" memories will be heavily mediated by photographs, videos and *Facebook* posts. The instant digital mediation of tourist activity lends further grist to the mill of those who believe that such experiences are essentially meaningless, brief encounters with a complex city which only locals really understand. Many of these locals (particularly the academics) despair about the resulting commodification of Oxford's history and identity into gift shops and over-priced museum cafes. But to write off and disparage tourism is to ignore so much of the vitality of contemporary Oxford. After all, what makes my own insider experience of the Radcliffe Camera any more valuable than that of the tourists who are kept safely outside? Aren't Oxford's students and scholars actually just types of tourists themselves, drawn to study or work in the city by an ideal of Oxford-ness and its historical associations with knowledge and power? As Leshu Torchin describes, the critical discussion of tourism is dominated by a binary opposition between authenticity and artificiality, rendering tourists as either frustrated seekers of the real or anchorless revellers in the unreal [1]. In this article, I will follow Torchin by granting tourists greater agency to negotiate between these two poles. This agency is demonstrated through an active remapping of space, positing tourists as co-creators of a theoretical "deep map" of the city, one which enables the navigation of "Oxford" in both the real and the imaginary senses.

As the concept of the "deep map" is relatively unknown outside academic discourse, and not yet fully explored even in a critical sense, it is well worth beginning with a discussion of the term and its origins. Most obviously, the modification of the noun "map" with the adjective "deep" highlights the thinness and two-dimensionality of traditional cartography. All maps are symbolic and selective, and when we consider that they have frequently been created by those with social and political power, they can be read as a means of ideological control over the citizens of a place. Hence politically-motivated groups such as the French Situationists of the 1950s and 1960s literally deconstructed (as in chopped to bits and remade) street maps of Paris according to their personal emotional experience of the city. The Situationists called this process "psychogeography" and were particularly interested in the relationship between the environment and human behavior [2]. The term "deep map" probably originated some years later in the distinctive travel writing of William Least Heat-Moon: it is the subtitle of his detailed and multi-layered account of a small patch of the Kansas grasslands *PrairyErth* [3]. Whilst Heat-Moon chooses to avoid a precise definition of his methodology, the literary critic Susan Naramore Maher has unpicked his mysterious style in the following terms:

> What distinguishes the deep map form from other place-based essays is its insistence on capturing a plethora of interconnected stories from a particular location, a distinctive place, and framing the landscape within this indeterminate complexity. Deep maps present many kinds of tales in an effort to capture the quintessence of place, but the place itself remains elusive and incompletely limned ([4], pp. 10–11).

From a literary and more broadly a humanities perspective then, the deep map is essentially a means of reconciling place and narrative. This is not simply a matter of geolocating literary or cinematic texts, although much interesting work is being done in this area (see for example Julia Hallam and Les Roberts' *Locating the Moving Image* [5]). In Heat-Moon's work, deep mapping requires a "plethora" of "stories" and "tales", and results in a destabilization of concrete space: place becomes "indeterminate" and "elusive".

Such a description of place might equally be applied to popular publications such as Tony Reeves' *Worldwide Guide to Movie Locations* [6], where real places are fragmented into multiple pieces of filmic narrative and precise filming locations. Whilst Reeves' expansive global project is too broad to qualify as deep mapping, there are plenty of more detailed publications which feel closer to Heat-Moon's obsessively localized methodology. Pop into any Oxford gift shop and you will almost certainly find at least one example, such as Bill Leonard's *The Oxford of Inspector Morse and Lewis* [7], a slim but dense volume designed to accompany walking tours of the city and filled (like *PrairyErth*) with simplehand-drawn maps. Here the plethora of stories are mostly fictional—summaries of key episodes which draw attention to place over and above plot—but there are also historical accounts of the city and its landmarks. Such guides and the omnipresent "movie maps" produced by tourist boards are fascinating collisions between narrative and place, and yet they have received little attention from the growing critical literature on the phenomenon of screen tourism. Les Roberts work on the movie maps of the city of Liverpool is a notable exception in this regard.

Roberts argues that, for a city such as Liverpool which is often used as a cheaper cinematic "stand-in" for other cities—principally London—the benefits of screen tourism can be obscure at best. In fact, active "destination marketing" using screen products may only serve to exacerbate historical inequalities between the South and North of England. My own focus upon Oxford presents a very different dynamic in terms of class and English regional identity, and although Oxford also "stands-in" for Hogwarts, the fantastical blending of real and imaginary locations presents no obvious barrier to pleasure for the Harry Potter tourist visiting the city. Therefore unlike Roberts' example of Liverpool where movie maps arguably represent a "*disembedding* of place, identity and cultural memory" ([8], p. 201), I want to propose that screen tourists in Oxford use movie maps, guides and online cartography to enable a seductive fantasy of *embodiment*, a means to temporarily inhabit a fictional hero by entering their world. This fantasy space is not quite the real Oxford, and it requires different tools to navigate. These tools can be thought of as a "deep map" in the sense that they are a space where narrative and place are indistinguishable.

A different branch of academic research has recently taken up the literary metaphor of deep mapping and used it as a means of building bridges between the humanities and the social and physical sciences. This is especially driven by the development of technology known as Geographical Information Systems (GIS), which offer sophisticated means of mapping data. In an effort to encourage the use of GIS outside of the social sciences, David Bodenhamer, Trevor Harris and John Corrigan have called for humanities scholars to "design and frame narratives about individual and collective human experience that are spatially contextualized", enabling the construction of "increasingly more complex maps…of the personalities, emotions, values and poetics, the visible and the invisible aspects of a place" ([9], pp. 172–74). Whilst the construction of a deep map of Oxford using GIS is beyond the scope of this article, I will incorporate and analyze geospatial data which is already publicly available, in particular photographs taken by tourists in Oxford and posted online. Such data has the crucial deep mapping characteristic of user-interactivity: the map as modified and constructed by its users. This is compatible with Maher's notion of "interconnected stories", and although Heat-Moon is credited as the sole author of *PrairyErth*, the book is constructed from a complex web of quotations and tales from local people, giving the sense that it grows organically out of the land itself. Whilst these geotagged images and their commentary emerge digitally rather than organically, there is no reason why they should not be just as evocative of lived experience. The digital mediascape is certainly a recent phenomenon compared to the much broader historical scope of Heat-Moon's book, but when considered as a conduit for historical information, social media can function as an enabler—or even an automatic generator—of deep maps, as I hope to demonstrate below.

2. Inspector Morse: "Sophocles did it!"

According to novelist Philip Pullman, Oxford is particularly suited to an imaginative engagement with place due to its long history of producing fantastical narratives. From *Alice's Adventures in Wonderland* (1865) to C. S. Lewis' Narnia tales, to the urtext of fantasy literature *The Lord of the Rings* (1954–1955), Pullman's Oxford is a city where "the real and the unreal jostle in the streets" ([10], p. i). We don't need to go as far as this playfully supernatural conceit to acknowledge that central Oxford with its cobblestone streets, narrow lanes and omnipresent gargoyles is the kind of landscape that inspired much of the medievalist iconography of fantasy literature including most recently J. K. Rowling's Harry Potter novels.

Of course there are more prosaic ways to account for Oxford's connection with fantasy literature. In order to become a fantasy novelist in the 19th and early 20th centuries you needed access to a store of ancient knowledge about mythology and legend, as well as an occupation providing sufficient time to write. So it is certainly no coincidence that Lewis Carroll, C. S. Lewis, J. R. R. Tolkien and Phillip Pullman all studied or taught at Oxford University. The University so dominates the centre of Oxford both physically and imaginatively that Oxford as a fictional setting carries unavoidable associations of intellectualism. Hollywood cinema has occasionally exported all-American characters to the University setting in order to create dramatic or humorous counterpoint. In *A Yank at Oxford* (1938) rugged Lee Sheridan (Robert Taylor) is prized for his physical prowess on the running track and rowing on the river, but his forthright masculinity gets him into trouble with married and flirtatious Elsa (Vivien Leigh). *Oxford Blues* (1984) also contrasts the physicality and cocky sex appeal of its star Rob Lowe with a parade of eloquent but spineless English undergrads. It is also notable for completely disregarding Oxford's geography in its construction of space. For a more recent example, consider *X-Men: First Class* (2011) in which the English Charles Xavier (James McAvoy) is a graduate of Oxford University with mutant powers. These powers make him the ultimate personification of Oxford intellectualism: an all-powerful telepathic brain and (eventually) a worthless incapacitated body. All these narratives are animated by a tension between Oxford intellectualism and American physicality. Whilst none of these films are likely to incite tourist demand individually, as part of a collective imaginary of the city they offer an enticing fantasy for visitors.

However, as visitors to Oxford quickly discover, the city's intellectual power and its associated spaces are certainly alluring, but they are often inaccessible. There is only very limited access to the inside of the Radcliffe Camera for those without a precious library card, and whilst walking around the city centre, college life with its dorms, dining halls and quads can often only be glimpsed through locked gates carrying signs that unambiguously state "Closed to Tourists". This is a rude reminder that Oxford's brand of intellectualism functions as an exclusive club, wonderful and empowering for those inside but opaque and impenetrable to those who are not. Behind closed doors, secrets lurk, and secrets lead to corruption. The corruption of Oxford's elite has certainly not gone unexplored in narratives set in the city. Indeed the abuse of intellectual power is an important theme of Oxford's most successful and popular television export: *Inspector Morse* (1987–2000). *Morse* is a very useful example of a media tourist driver for several reasons. Firstly, the huge international success of the original and follow up series mean that Morse is a character well known to audiences all over the world. The show's long-form structure—running at two hours including advert breaks—was unusual for British television when it was first transmitted in the late 1980s, but meant that overseas it could be sold as a "miniseries" or even as a sequence of one-off TV movies. At its peak, the show's global audience was estimated to be around 1 billion viewers in over 200 countries [11]. Secondly, just as important as the series' international reach is its strong local flavor. The show makes extensive use of location shooting in and around Oxford, and this is an important element of its high production values and status as "quality television" [12]. *Inspector Morse*'s convoluted murder mystery plotlines typically rely upon the city's eccentricities, such as "The Wolvercote Tongue" (first transmitted on Christmas Day in 1988) which has a self-reflexive interest in Oxford tourism as criminal suspicion falls upon tourists, tour guides and even employees of the Ashmolean Museum. Finally, as Steijn Reijnders has pointed out, detective narratives have a topophilic nature, and a particular affinity with tourist activity, in that a fan can literally "follow in the footsteps of his or her beloved inspector, criss-crossing the local community, looking for signs and clues" ([13], p. 177).

As Reijnders suggests, the intense relationship here is not only between narrative and place, but between place and character, the "beloved" detective who is inseparable from the place he or she investigates. By exploring the visited location in detail, sometimes going back over the same spots many times, tourists can find signs which accumulate into a kind of cartography of character: personality as action which takes place on a map. The key elements of Morse's character all locate him spatially within Oxford and provide suggestive clues for his tourist fans. What separates Morse most clearly from the majority of policeman from British crime drama is that he is presented as a cultured intellectual. In this respect he has more in common with earlier sleuths Sherlock Holmes and Hercule Poirot than his contemporaries, such as John Thaw's previous role in the genre, the tough-talking DI Jack Regan from *The Sweeney* (1975–1978). Morse inhabits Oxford intellectualism in both literal and symbolic ways. In the most literal sense, Morse was himself an Oxford undergraduate, although he failed his final degree under mysterious circumstances. This trope places the character in an interestingly ambivalent relationship with the city's intellectual elite. The show's plotlines often depict this same elite as corrupt in murderous ways, setting up a problematic relationship between knowledge and power which is perhaps at the heart of the show's appeal. To give just one example: in "The Last Enemy" (first broadcast in January 1989), a body is found in the Oxford canal, leading Morse to investigate the fellows of a fictional composite college called "Beaumont". It transpires that rivalry for a senior academic post led to the first murder, committed by another Professor, and a further two killings also take place amongst the college's academic staff. In Morse's Oxford, intellectualism gives access to the secretive spaces of the colleges where corruption festers.

Despite this ambivalence around the use and control of knowledge, the detective genre requires knowledge to be gradually gained and acquired throughout the narrative arc. This means that Morse will always be simultaneously enthralled and appalled by Oxford intellectualism; he needs knowledge to do his job, but what he learns is frequently disturbing to the status quo. How Morse gains knowledge is both a physical, located process—he drives (in his vintage Jaguar), walks or runs around Oxford's landmarks—and an intellectual one: he ponders, thinks and researches like an academic, using theories where evidence is absent or misleading. For example, in his first television film "The Dead of Jericho", first broadcast in January 1987, Morse attempts to solve a case involving the death of a maternal figure with both a classical (and psychoanalytical) conclusion that "Sophocles did it!" This turns out to be a red herring, but it is a misfire which almost makes sense thanks to the rarefied Oxford setting. The same episode also introduces Morse's passion for classical music, particularly of the choral or operatic kind. As Pierre Bordieau described, an affiliation with classical music or opera is a particularly strong bearer of cultural capital in Western society due to its perceived inaccessibility or difficultness [14], so Morse's extensive knowledge of the subject marks him as a middle-class traditionalist. Unlike visual art, novels and films, music does not directly represent landscape or place and it can therefore be thought of as the most abstract and least spatial of the art forms. However, music has a particularly strong connection with emotion and memory, and famous pieces of music build up webs of cultural and personal associations. These associations and memories are often spatially located, so it is certainly not impossible to incorporate music into a multi-media deep map of Morse's Oxford, especially given that Morse's use of music within the diegesis is especially localized. Playing records is his major pastime whilst off duty, in fact a brief shot of a turntable with diegetic classical music is usually enough to establish that Morse is at home, but he also sings in a choir which rehearses in recognizable tourist landmarks such as the Holywell Music Rooms.

Music is vital to the characterization of Morse and therefore the show itself, so the producers of one particular guide to Morse's Oxford showed impressive ingenuity when they included a bonus CD of soundtrack music stapled to the front cover [15]. The potential uses of this music in terms of memory work are very interesting: it could function either as a powerful and exciting reminder of the show whilst planning a trip, or a pleasurable *aide memoire* of both the show and the trip it inspired in years to come. Doubtless some tourists even bring this type of musical accompaniment with them on portable devices to listen to whilst travelling or even whilst walking round the streets in Morse's footsteps. In a similar vein, Gundolf Graml's work on *Sound of Music* tourism notes the "kinaesthetic experience" of travelling through the Austrian landscape on open-topped tour buses which play music from the film to blur the boundary between filming location and narrative setting ([16], p. 152). Such multimedia approaches to locating screen narratives pre-figure the boundless opportunities presented by the uptake of social media and smartphone technology. Those who are now researching a trip to Oxford and want to explore Morse locations have a multitude of options: from traditional guidebooks, to video guides on *YouTube* [17], to maps constructed using adaptable mapping technology. Figure 1 shows one example of a user-modified *Google Map* of Morse's Oxford including places described in Colin Dexter's original novels, shooting locations from the television series and even walking routes which, like Reijnders detective tourists "criss-cross" the city centre [18]. When used with an internet enabled smartphone this map and others like it are clearly capable of displaying the tourist's real location within Morse's fictional world.

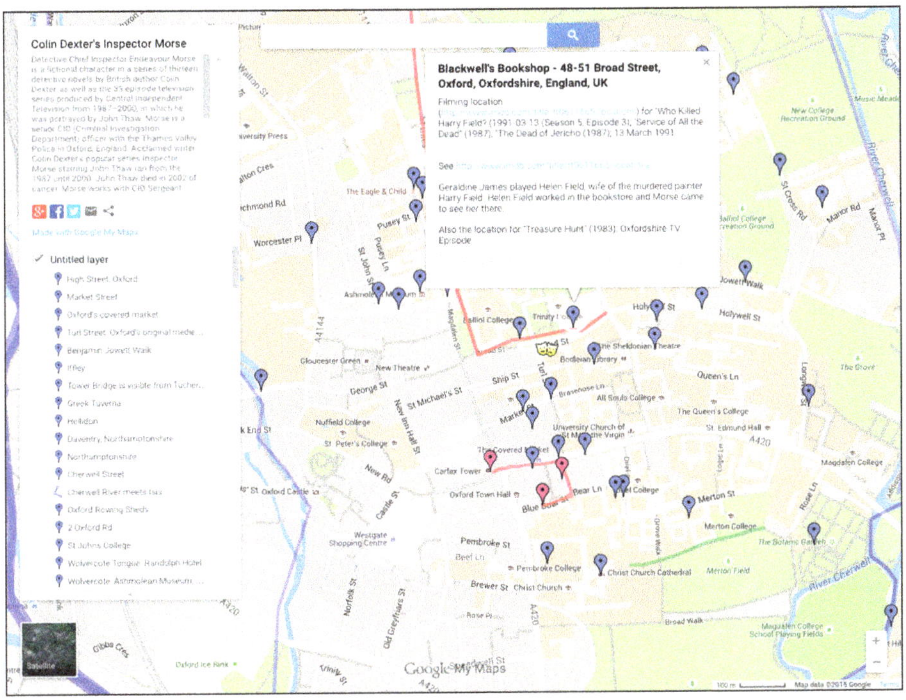

Figure 1. User-modified *Google Map* of "Colin Dexter's Inspector Morse" [18]. ©2015 Google.

Figure 2 is taken from *Flickr*, a photo sharing service used by keen amateur and semi-professional photographers [19]. Because most images taken with modern devices are automatically "geo-tagged" (*i.e.*, marked with geographical data indicating where there were taken), *Flickr* can present the uploaded and publicly-shared photographs in the form of a searchable map. Each of the pink dots on the map can be selected to display a preview of the image, which when clicked is opened in higher definition and with comments by users and fans. In this example the possibilities of the internet to enable or even automatically generate deep maps begins to become clear. This map captures images which were inspired by stories; written as novels, adapted onto television screens and then played and replayed around the world. This type of map certainly fits both

Heat-Moon and Bodenhamer *et al.*'s definitions of what a deep map is and how it may enhance our understanding of place and narrative. The images it relays are in themselves fragments of narratives—of tourists' experiences and fans engagement with fictional heroes—as I will explore in more detail in the following section.

Figure 2. *Flickr*'s searchable map of images uploaded in Oxford [19]. © 2012 Digital Globe.

3. "Any of You Harry Potter Fans Recognize Where I'm at?"

Just as with *Inspector Morse*, the huge global success of the *Harry Potter* franchise (2001–2011) is well documented, as is the powerful draw of the films for international tourists. Chieko Iwashita's survey of Japanese tourists visiting the UK in 2002 found that over a third of respondents cited *Harry Potter and the Philosopher's Stone* (2001) as the film which had most increased their interest in visiting the UK, and it seems certain that this effect has only intensified as further films in the series were released [20]. However the franchise's fantastical blending of many different locations with digital effects work makes the location of tourist demand diffuse throughout the UK. Whilst a few "real" locations are named and hence visited (e.g., King's Cross Station), in the Harry Potter films the landscape of the UK speeds by like a kaleidoscope of spectacular valleys, towering turrets and dusty libraries. Whilst there are certainly keen Harry Potter fans who engage with targeted tourism over large distances, it is important to acknowledge that the lure of visiting a real life bit of Hogwarts in Oxford is generally a part of a larger set of forces. The long queues outside the well-known Potter locations of Christchurch College or the Bodleian Divinity School are made up of much younger fans than those of *Inspector Morse*, with many still in their teens and therefore unable to make the same types of decisions to travel available to older tourists. Therefore Potter fans come with their families, or in larger school and college coach tour size groups. Such family or school tours would probably have happened even without the additional draw of Harry Potter.

Tourists are complex individuals with a myriad of motivations to travel, and this is one of the primary reasons why establishing the precise impact of film and television products on tourist activity is extremely difficult. One empirical study which rose to this challenge interviewed tourists in Oxford in 2009 and found "it was only 28% true to say that but for screen portrayals they would not have visited Oxford. In other words, it was 72% true to say that they would have visited Oxford anyway, even if they had never seen Oxford or the UK portrayed on screen." ([21], p. 207). However the same survey also found that 70% of visitors to Oxford *had* been influenced by screen products *to some degree*, which given the size of tourist numbers over the summer is still an extremely significant factor. It is no surprise then to find that the choice of destinations within Oxford has clearly been skewed by Harry Potter tourism. The Dean of Christ Church Cathedral stated in 2012 that visitor numbers had grown significantly over the previous decade, and that the average age of visitors was now much lower than before the Potter franchise was released [22]. Even the Bodleian Library acknowledged the power of Potter by including J. K. Rowling's annotated edition of *Harry Potter and The Philosopher's Stone* (1997) in their "Magical Books" exhibition of 2013 alongside Tolkien's maps and drawings. The multiple motivations of screen tourists are further proof that even deeper maps of the city are required than those which incorporate screen narratives and filming locations. Shopping, cinema-going and café culture are also on the map for younger tourists, just as museums and galleries are vital for more mature visitors.

Due to the complexity of their production processes and the fantastical nature of their mise-en-scène, the *Harry Potter* films place greater physical and imaginative demands upon their tourist fans than those of other more localized products, such as *Inspector Morse*. Imaginative work is required because the physical filming locations used are almost always "playing" an imaginary other place. This means that they may look very different "in real life", as they were expensively dressed or digitally manipulated before making it onto the screen. Christina Lee's study of Harry Potter tourism argues that the potential disparity between the real and the fictional places is reduced when the locations are "historically authentic", such as the dining hall of Christ Church College; which is a functioning mess hall for a great educational institution both on and off screen ([23], p. 58). Nonetheless, there remains a layering effect where the pre-existing cultural context of the real location overlaps with the fictional universe. Thus the deep map of Oxford for Harry Potter tourists is complex and multi-layered, comprising lived experience of physical Oxford, much-anticipated but uncertain encounters with the film location itself, and an imaginative interplay between the location and memories of the film's spectacular imagery. Much of this cognitive work would be difficult to evidence were it not for the fact that the younger average age of the Harry Potter fan means that they are very likely to be heavy users of social media, especially when travelling.

It is plain to those who have witnessed younger tourists in a city such as Oxford that many of them are not just gazing at landmarks, they are also taking photos—often of themselves—with their smart phones, pausing to write brief tags or status updates, and then instantly posting them online. The UK's media regulator Ofcom carries out an annual survey of media use and attitudes, which in 2015 confirmed that 90% of UK respondents aged 16 to 24 use social media at least once a week, compared to just 44% of those aged 55 to 64. In addition this use is far more likely to be mobile for younger users, who go online via smartphones almost as much as computers or laptops [24]. For international tourists, historically high data roaming charges used to limit internet access abroad, but within the EU this price restriction is less significant since legislative changes in 2014, and in any case densely-populated city centers such as Oxford have a high concentration of free Wi-Fi hotspots for tourists to use when posting online. These rapid technological changes have certainly contributed towards a generational shift in attitudes concerning the public/private divide, especially when it comes to photographs. As social media observer Jacob Silverman puts it: "'Pics or it didn't happen'—that is the populist mantra of the social networking age. Show us what you did, so that we may believe and validate it" [25]. Naturally then, for unusual activity such as visiting the shooting location of your favourite film, instantly proving that "it really happened" via images on social media is increasingly important to younger tourists.

The implications of this behavior for the tourist experience as a whole are wide-reaching. As John Urry has noted, taking photos has long been central to the holiday experience as a means of capturing "the tourist gaze" [26], but prior to the uptake of social media the primary use of these images was one of private memory work, or of building narratives to be relayed at a later date to friends and family once back home from a trip. The instantaneous relay of these experiences to one's "friends" or even to the wider public adds a new dimension of performativity, which is particularly well-suited to the photography habits of the screen tourist. Stefan Roesch's survey of *The Experiences of Film Location Tourists* details many variants of playful image making amongst movie fans including "shot re-creations", "filmic re-enactments" and even "miniature positioning": for example placing toy figurines of *Star Wars* characters so that real desert shooting locations form a cinematic backdrop ([27], pp. 159–80). The many thousands of images of Harry Potter fans visiting Oxford locations which are publicly-available on *Instagram* (a photo sharing service particularly popular amongst younger social media users) tend to be rather more spontaneous and ad hoc in nature, but display a similar passion for embodiment within the cinematic universe.

I will discuss just three representative examples of images posted on *Instagram* in July 2015. In the first, a young woman combines smiling images of herself standing on the stairs of the Divinity School and in Christ Church Dining Hall and comments "Happier? Impossible!" In the second, a pair of young women pose in the same Dining Hall, grinning and pointing at their handy visual prop: the DVD of *Harry Potter and the Chamber of Secrets* (2002). And finally in the third a young man student stands in the cloisters of New College next to an ancient oak tree recognizable from the franchise and poses the challenge: "Any of you Harry Potter fans recognize where I'm at?" His friend replies: "The real Hogwarts" [28–30]. Figure 3 combines these three images, all of which are publicly available on the *Instagram* service, however the text and comments have been omitted to protect other users' identities. All three examples display a genuine and infectious excitement on behalf of the screen tourists pictured: big goofy smiles like these are impossible to fake for the camera. Clearly these moments of collision between reality and fantasy are meaningful and pleasurable for a great many people, even including academic researchers, as Sue Beeton's enthusiastic experience of "The Wizarding World of Harry Potter" confirms [31]. But the fact of their posting on the *Instagram* service turns these images into something other than private memories: they function more like elements of a conversation between friends and sometimes like-minded strangers. The participants in the larger global conversations sparked by the images are fan communities who can find the images because they are (hash)tagged with key words such as "#harrypotter" or "#hogwartsismyhome". They are also localized, both via the tag "#oxford" and because, as with the *Flickr* photos taken by fans of *Inspector Morse*, they can be presented in the form of a map. Therefore *Instagram* and its fellow image-sharing services together posit a global map of experience which becomes increasingly deeper as the services mature and accumulate. There is much further research to be done on this subject in order to uncover the riches of these visual sources.

Figure 3. *Instagram* photos posted by Harry Potter fans in Oxford [28–30].

4. Conclusions

All maps are an argument about space, and all maps have a purpose. Most obviously, maps help us to navigate unfamiliar terrain and avoid getting lost. They lead us to places which we want to see and help us find places which fulfill our basic needs. They may also demarcate territories and reinforce political power. But as historian of cartography Jerry Brotton argues, maps also fulfill a deeper imaginative function, as they are "always images of elsewhere, imaginatively transporting their viewers to faraway, unknown places" ([32], p. 15). Considering the map as an instrument of fantasy brings us back to the particular focus of this article: international tourists to Oxford who are (at least partly) motivated to travel by cultural texts. When the must-see object is less a building or place than a fictional character that inhabits it, a traditional map is not fit for purpose. What is required is a deeper engagement between place and narrative, a space where stories are just as important as history, and history has many of the best stories. To return to the author's working environment discussed at the start of this article, when eating lunch on the steps of the Radcliffe Camera in Oxford there are ample opportunities for eavesdropping upon the many walking tours which pause by the building to contemplate its famous curves. Listening for more than a few minutes the same stories get repeated over and over again, echoing round the walls of Radcliffe Square: one favourite with tourists concerns a duck which flew out of the foundations of All Souls College in 1437, prompting a bizarre college tradition involving a moonlight procession led by a "Lord Mallard".

The telling and retelling of these tall tales highlights the extent to which the tourist experience is made up of stories. Stories provide insight into the place visited, and may be communicated to the tourist via a bewildering array of media. Screen narratives provide compelling characters who inhabit particular places so completely that to visit the place is to feel a deep connection between reality and fantasy. The tools used (and increasingly created) by tourists in order to navigate these imaginary places may be thought of as "deep maps", in the sense that they "capture a plethora of interconnected stories from a particular location" [4]. They are guides not just to finding filming locations but also ways to understand "the personalities, emotions, values and poetics, the visible and the invisible aspects of a place" [9]. They are heavily mediated and re-mediated, and yet they clearly provide meaningful emotional experiences for visitors who may be searching for clues like Inspector Morse or marveling at the magical places touched by Harry Potter himself. In this sense, screen tourist activity participates in a collective deep mapping of destinations such as Oxford, enabling all visitors, be they day-trippers or visiting scholars, to navigate the relationship between narrative and place.

Acknowledgments

The ideas in this article were first explored in a conference paper given at the ECREA Conference at the University of Lund in 2013. My thanks go to the organizers of this event and the participants who provided useful feedback. A similar debt is due to those attending the recent symposium "Developing Film Tourism: Theory and Practice" held at the University of Glasgow in May 2015. In particular I want to acknowledge the influence of an instructive paper given in Glasgow by Meghann Ormond on embodiment in medical tourism. Finally, many thanks to *Instagram* users Misako Kawamura, Matthew Holt and Carmen Temmler for giving permission to use their images in this article.

Conflicts of Interest

The author declares no conflict of interest.

References

1. Leshu Torchin. “Location, Location, Location: The Destination of the Manhattan TV Tour.” *Tourist Studies* 2 (2002): 247–66.
2. Simon Ford. *The Situationist International: A User’s Guide*. London: Black Dog, 2005.
3. William Least Heat-Moon. *Prairyerth: A Deep Map*. London: Deutsch, 1991.
4. Susan Naramore Mayer. *Deep Map Country: Literary Cartography of the Great Plains*. Lincoln: University of Nebraska Press, 2014.
5. Julia Hallam, and Les Roberts. *Locating the Moving Image: New Approaches to Film and Place*. Bloomington: Indiana University Press, 2013.
6. Tony Reeves. *The Worldwide Guide to Movie Locations*, 3rd ed. London: Titan Books, 2006.
7. Bill Leonard. *The Oxford of Inspector Morse and Lewis*. Stroud: Tempus, 2008.
8. Les Roberts. “Projecting Place: Location Mapping, Consumption and Cinematic Tourism.” In *The City and the Moving Image: Urban Projections*. Edited by Richard Koeck and Les Roberts. Basingstoke: Palgrave Macmillan, 2010, pp. 183–201.
9. David J. Bodenhamer, Trevor M. Harris, and John Corrigan. “Deep Mapping and the Spatial Humanities.” *International Journal of Humanities and Arts Computing* 7 (2013): 170–75.
10. Phillip Pullman. *Lyra’s Oxford*. London: Corgi Books, 2003.
11. William Gallagher. “Morse’s Last Grumble.” *BBC News*, 14 November 2002. Available online: http://news.bbc.co.uk/1/hi/entertainment/1022958.stm (accessed on 27 May 2015).
12. Lyn Thomas. “In Love with Inspector Morse: Feminist Subculture and Quality Television.” *Feminist Review* 51 (1995): 1–25.
13. Stijn Reijnders. “Watching the Detectives: Inside the Guilty Landscapes of Inspector Morse, Baantjer and Wallander.” *European Journal of Communication* 24 (2009): 165–81.
14. Pierre Bourdieu. *Distinction: A Social Critique of the Judgement of Taste*. London: Routledge & Kegan Paul, 1984.
15. Annie Bullen. *Morse in Oxford*. Andover: Pitkin, 2009.
16. Gundolf Graml. “(Re)mapping the Nation: Sound of Music Tourism and National Identity in Austria, ca 2000 CE.” *Tourist Studies* 4 (2004): 137–59.
17. Studio 8 Video Production. “Inspector Morse’s Oxford: Introduction.” 2010. Available online: https://www.youtube.com/watch?v=QrexVyBDn1o (accessed on 2 July 2015).
18. Google Maps. “Colin Dexter’s Inspector Morse.” 2015. Available online: https://www.google.com/maps/d/u/0/viewer?mid=zEiHoRM0-99k.kD_DPkIlS_0c (accessed on 2 July 2015).
19. Flickr. “World Map.” 2015. Available online: https://www.flickr.com/photos/tags/inspector%20morse/map?&fLat=51.7507&fLon=-1.257&zl=16 (accessed on 16 July 2015).
20. Chieko Iwashita. “Roles of Films and Television Dramas in International Tourism: The Case of Japanese Tourists to the UK.” *Journal of Travel & Tourism Marketing* 24 (2008): 139–51.
21. Anita Fernandez Young, and Robert Young. “Measuring the Effects of Film and Television on Tourism to Screen Locations: A Theoretical and Empirical Perspective.” *Journal of Travel & Tourism Marketing* 24 (2008): 195–212.

22. BBC. "Harry Potter Fans Boost Oxford Christ Church Cathedral." *BBC News*, 25 March 2012. Available online: http://www.bbc.co.uk/news/uk-england-oxfordshire-17434129 (accessed on 28 February 2015).
23. Christina Lee. "'Have Magic, Will Travel': Tourism and Harry Potter's United (Magical) Kingdom." *Tourist Studies* 12 (2012): 52–69.
24. Ofcom. "Adults Media Use and Attitudes Report 2015." 2015. Available online: http://stakeholders.ofcom.org.uk/market-data-research/other/research-publications/adults/media-lit-10years/ (accessed on 6 June 2015).
25. Jacob Silvermann. "'Pics or it didn't happen': The Mantra of the Instagram Era." *The Guardian,* 26 February 2015. Available online: http://www.theguardian.com/news/2015/feb/26/pics-or-it-didnt-happen-mantra-instagram-era-facebook-twitter (accessed on 25 March 2015).
26. John Urry, and Jonas Larsen. *The Tourist Gaze 3.0. Theory, Culture & Society*. London: SAGE, 2011.
27. Stefan Roesch. *The Experiences of Film Location Tourists*. Bristol: Channel View, 2009.
28. Carmen Temmler. "Happier? Impossible!" 2015. Available online: https://instagram.com/p/5X4JasCUEQ/ (accessed on 21 July 2015).
29. Misako Kawamura. "#canteen of #HarryPotter." 2015. Available online: https://instagram.com/p/5ZotzKpfqT/ (accessed on 21 July 2015).
30. Matthew Holt. "Any of you Harry Potter fans recognize where I'm at?" 2015. Available online: https://instagram.com/p/5W8tiPL15t/ (accessed on 21 July 2015).
31. Sue Beeton. *Travel, Tourism and the Moving Image*. Bristol: Channel View Publications, 2015.
32. Jerry Brotton. *A History of the World in Twelve Maps*. London: Allen Lane, 2012.

Regular Routes: Deep Mapping a Performative Counterpractice for the Daily Commute [1]

Laura Bissell and David Overend

Abstract: This article offers a textual "deep map" of a series of experimental commutes undertaken in the west of Scotland in 2014. Recent developments in the field of transport studies have reconceived travel time as a far richer cultural experience than in previously utilitarian and economic approaches to the "problem" of commuting. Understanding their own commutes in these terms—as spaces of creativity, productivity and transformation—the authors trace the development of a performative "counterpractice" for their daily journeys between home and work. Deep mapping—as a form of "theory-informed story-telling"—is employed as a productive strategy to document this reimagination of ostensibly quotidian and functional travel. Importantly, this particular stage of the project is not presented as an end-point. Striving to develop an ongoing creative engagement with landscape, the authors continue this exploratory mobile research by connecting to other commuters' journeys, and proposing a series of "strategies" for reimagining the daily commute; a list of prompts for future action within the routines and spaces of commuting. A range of alternative approaches to commuting are offered here to anyone who regularly travels to and from work to employ or develop as they wish, extending the mapping process to other routes and contexts.

Reprinted from *Humanities*. Cite as: Bissell, L.; Overend, D. Regular Routes: Deep Mapping a Performative Counterpractice for the Daily Commute. *Humanities* **2015**, *4*, 476–499.

1. Introduction

In 2011, we set up Making Routes; a network and online resource for researchers and practitioners working with mobilities in contemporary performance [2]. We organised a symposium at the Arches arts centre in Glasgow [3], and ran a series of events including Phil Smith's *Misguided in Ayr* [4]. Our interest in journeys continued independently, with the occasional opportunity to exchange ideas and discuss our various projects. We think that the idea of working together was always there, but it was not until September 2013 that we decided to collaborate on a research project. At the beginning, neither of us was clear on what that might be. We thought we might travel somewhere together, or make a performance about journeys. We started without a destination in mind, which is probably the best way to start.

Lacking a clear direction, we found it difficult to make time to work together, and with both of our lives including full-time lecturing posts and long commutes, we were initially anxious about when we would be able to embark on the journeys we wanted to take: Laura travels by car, ferry and train every day from Innellan on the West Coast of Scotland to the Royal Conservatoire of Scotland (RCS) in Glasgow; David drives from Glasgow to the University of the West of Scotland's (UWS)

[1] An early version of this article was presented as a keynote paper at the British Sociological Association Regional Postgraduate Event at Glasgow Caledonian University, UK, 13 June 2014 [1].

Ayr campus. It was for a practical reason, then, that the first stage of this project was for us to accompany each other on our commutes, and it was during these specific and purposeful journeys, and the subsequent meal at each other's homes, that we started to think critically about our daily, repetitive mobility. Laura kept a research diary of these initial conversations:

> Tuesday 24th September 2013
>
> 17:25 train, 18:20 boat, arrive home at 19:00
>
> The train is busy. David and I travel back to my house together to talk about our project. David tells me that his commute is very different as his is solo in a car while mine is mainly made up of public transport surrounded by other bodies and has four different stages. We talk about lots of things that are going on in our lives (work, relationships, performances we have seen) but we don't talk much about the project until we are at my house and having some dinner that I have prepared the night before. We come up with some areas to research and both agree that doing each other's commutes is an important part of this process. We arrange for me to do David's commute on the 7th October and through this conversation I see my journey through David's eyes. I drive him to the ferry at 20:45 and David travels back to Glasgow alone [5].

Our discussions around this project encouraged us to consider documenting our commutes as times and spaces of possibility and creative resistance—a way of reimagining aspects of the working day. This article focuses on the first stage of this process—an attempt to develop a "deep map" of our experiences-so-far.

Commuting has often been dismissed as a "desensitising" experience in which Sennett identifies a "disconnection from space as the body moves passively" through urban roadscapes ([6], p. 18). Likewise, Jean Baudrillard refers to the "effortless mobility" of car journeys ([7], p. 66); and Michel de Certeau points out the "immobility" engendered by train travel ([8], p. 111). In the field of transport studies and policy, the perception of commuting as a "problem" has led to an assumption that travel time is inherently unproductive ([9], p. 81). As Patricia Mokhtarian points out, as a result "virtually all of our policies, planning, and models are predicated on the assumption that travel is a disutility to be minimised" ([10], p. 93). However, others have recognised greater complexity in the experience of contemporary mobilities, which involves an uneasy combination of "experimentation and danger, possibility and risk" ([11], p. x). Identifying this complexity in our everyday journeys, we are interested in capturing our commutes as far more active and embodied experiences than the unsatisfactory interpretations of many critical representations, and transport policies. Like Tim Edensor, we distrust the popular perception that the commuter is "a frustrated, passive and bored figure, patiently suffering the anomic tedium of the monotonous or disrupted journey" ([12], p. 189). Instead, we are interested in the potential of travel time as a "gift" that is "a desirable time for many people" who value the opportunity to transition between modes of living and working, and to take "time out" as a retreat from external demands ([9], p. 88).

In the sections that follow, we present a "deep map" of our explorations into our daily commutes as fertile ground for creativity, productivity and transformation. We present a series of autoethnographic narratives, documenting our alternative commutes, which we undertook initially by walking, and then by cycling, swimming and boating the route between our homes and our places of work. Together, this interconnected web of personal account, anecdote, critical reflection and reminiscence, offers a textual map of these regular routes. Mike Pearson and Michael Shanks identify the potential of such a map—a form of "theory-informed story-telling" ([13], p. 16)—to contain juxtapositions and interpenetrations of the historical and the contemporary, the political and the poetic, the factual and the fictional, the discursive and the sensual; the conflation of oral testimony, anthology, memoir, biography, natural history and everything you might ever want to say about a place ([14], pp. 64–65).

This approach "attempts to record and represent the grain and patina of a *location*" (our italics) ([13], p. 15). However, for this project, we are interested in deep mapping a *route*. This process attempts not only to map—to record, document, represent—but also to engage with our commutes actively through a performative engagement with landscape; mobilising a variety of experiences that work against the "desensitising" systems of contemporary mobile life. Our approach draws on the "mobility turn" in the field of performance studies, and recent research into "a wealth of performances and related cultural practices [that] have been, and continue to be, actively engaged in imagining, exploring, revealing and challenging experiences of being in transit" [15]; ([16], p. 1). Fiona Wilkie recognises the affective quality of performance in this context, as "performance not only responds to but can also produce mobilities, reshaping existing models and engendering new, alternative possibilities for movement" ([16], p. 2).

Importantly, we do not see this particular iteration of our project as an end-point. As such, we have aimed to leave our discursive map open ended, developing a series of experiments and "strategies" for reimagining the daily commute and extending our mapping project beyond our own journeys. These take the form of a postcard project on the Gourock to Dunoon ferry route (see Figure 1) and a list of instructions, which we offer here to anyone who regularly travels to and from work to employ or develop as they wish. Similar to Carl Lavery's *25 Instructions for Performance in Cities* [17], or Phil Smith's Mythogeographical "Toolbag of Actions and Notions" ([18], pp. 144–75), these strategies are "there to be given life by their readers", not as endpoints of our own explorations ([18], p. 9). Ultimately, they are offered as prompts to performative action within the routines and spaces of commuting. In recording and representing such activities, a performative document will emerge that extends our mapping process beyond the scope of this article to engage with other forms and methods of mapping everyday journeys. Alongside the deep map presented here, we continue to gather stories and images of experimental commuting journeys on our website: www.makingroutes.org [2].

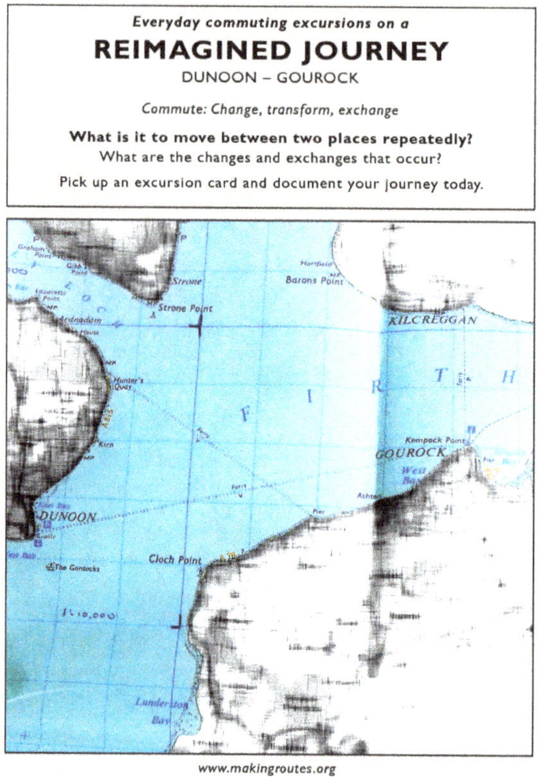

Figure 1. Everyday Commuting Excursions on a Reimagined Journey. Designed by Rachel O'Neill.

2. Performing Striated Space

Our mapping process is influenced by concepts of mobility drawn from nomadic theory: the condition of nomadism; and an attempt to incorporate a nomadic form of mobility into our regular journeys. Drawing on the "nomadology" of Deleuze and Guattari [19], and Rosi Braidotti [20], we aim to document a performative "counterpractice", learning more about our own commuting routes by considering alternative ways of traveling between two sites, and examining our exiting daily journeys through the lens of these key theorists of contemporary mobilities. Before discussing our own experiences, it will therefore be valuable to consider the potential for, and limitations of, a nomadic form of commuting.

As we have discussed, commuting is often portrayed as a problematically limited, constrained and regulated mobile practice. In contrast, Pearson sees the nomad as an aspirational figure, "cut free of roots, bonds and fixed identities" ([21], p. 20). Nomadism offers a valuable counterpoint to commuting, then, but it is a concept that we are anxious about adopting too readily without confronting some of the potential problems with its use. As Tim Creswell points out, although the metaphor lends itself to an antiessentialist discourse that emphasises "the importance of *becoming*" over that of destinations, the use of the nomad in Western intellectual thought is essentially "a form of imaginative neo-colonialism" which uses the non-Western other to project romanticised notions of mobility against the sedentary power relations of the State ([22], p. 47, 54). Embracing the concept of the nomad runs the risk of ignoring "who travels and why" ([23], p. 361).

These questions are particularly pertinent to a discussion of commuting, as a mobile practice that is constituted through significant social and economic hierarchies. Gil Viry *et al.* identify a range of factors that lead to inequality in contemporary commuting practices that is predicated on access to certain forms of mobility. These include not owning a car, residential areas with poor access to public transport, and "weak temporal or organisational resources to handle projects that require travel" ([24], p. 121). These factors contribute to an important shift in the way we acquire and build capital in a mobile world, which John Urry refers to as "network capital" ([25], p. 197). The ability to move (or to choose not to move) now constitutes a "major source of advantage" in a society that relies, more than ever before, on systems and processes of mobility ([25], p. 52).

Deleuze and Guattari's discussion of nomadic forms of mobility provides a valuable aspirational counterpoint to this heavily systematised, regulated and unequal spaces traversed through commuting. Unlike the "striated" space of roads, ferry routes and trainlines that we use to transport ourselves between work and home, nomad space is *smooth*, defined by "the variability, the polyvocality of directions" ([19], p. 382). Moving within vast, open spaces such as deserts, tundras and steppes, nomads make paths, moving from point to point through the landscape. However, as opposed to sedentary societies, the points of a nomad journey are "strictly subordinated to the paths they determine", rather than the other way round ([19], p. 380). Importantly, although commuting does not fit easily within this model, this is not to say that the commute precludes a nomadic relationship with the environment:

> Even the most striated city gives rise to smooth spaces [...] Movements, speed and slowness, are sometimes enough to reconstruct a smooth space. Of course, smooth spaces are not in themselves liberatory. But the struggle is changed or displaced in them, and life reconstitutes its stakes, confronts new obstacles, invents new places, switches adversaries ([19], p. 500).

Bearing in mind Deleuze and Guattari's warning that we should "never believe that a smooth space will suffice to save us", we recognise that our creative explorations of our own commutes are significantly limited in their ability to address the network capital that we require to go "off route". We are nonetheless setting out on a personal journey to "confront new obstacles" and "reconstitute the stakes" of regular journeys which have occasionally left us feeling "desensitised" to our own bodily experience of the landscapes that we move through. In this way, the example of the nomad offers us a new way of thinking about and experiencing our commutes, rather than reconstituting the systems that contain them and the socio-economic context that we travel within. Deleuze and Guattari's notion of *reconstitution* resonates with Braidotti's concept of nomadism as "a myth, or a political fiction, that allows us to think through and move across established categories and levels of experience" ([20], p. 26). Braidotti advocates "nomadic shifts"; "a performative metaphor that allows for otherwise unlikely encounters and unsuspected sources of interaction, experience and knowledge" ([20], p. 27). In these shifts, our everyday practices are opened up to the creative processes of *becoming*, offering an "acute awareness of the nonfixity of boundaries" and "the intense desire to go trespassing, transgressing" ([20], p. 66).

Critical appropriations of nomadism have acquired a great deal of currency in postmodern accounts of mobility and it is easy to see how they suit our purpose here. These concepts offer an appealing invitation to the disgruntled daily commuter to leave the striated space of the commute and go "trespassing" and "transgressing". The details of the specific journey must be taken into account, although the tools we offer can be applied to any commute in an attempt to enact "nomadic shifts". What mode of transport is employed? Is it public or private? What is the usual pace of this journey? Is the journey open to disruption? Wilkie identifies that discourses of walking practices such as pace, value, autonomy and agency can also be applied to road travel but questions whether "motorvating" actually does open up "previously unavailable opportunities for freedom, escape and subversion" ([16], p. 85) or whether the "democratisation" of road travel has remained mythic for those without network capital.

As pointed out by Janet Wolff, the "suggestion of free and equal mobility is itself a deception, since we don't all have the same access to the road" ([26], p. 235). As a result, Cresswell has been challenging the "romanticisation of the nomad as the geographic metaphor *par excellence*: for some time ([22], pp. 42–54); ([23], p. 361). Embracing the metaphor of "generalised heroic figures resisting the confines of a disciplinary society" (itself a specifically postmodern appropriation of a previously vilified cultural figure), is to "gloss over the real differences in power that exist between the theorist and the source domain of the metaphors of mobility" ([23], pp. 378–79). It is important to point out, therefore, that we are not claiming that our creative experiments in the spaces of our daily commutes could be described as *nomadic*, or indeed that such a thing as nomadic commuting could ever exist, let alone be captured on a map. Rather, we are using "nomadology", and the *creative becoming* the myth of the nomad encapsulates, to prompt an alternative way of thinking about quotidian and functional journeys that prioritises creativity and imagination over an assumption of disutility within striated space. We aim to "mobilise metaphors of mobility" by exploring different ways of engaging with the landscapes that we move through between home and work ([22], p. 56). By temporarily departing from our usual modes of transport, we intend to re-perform our own repetitive forms of mobility and reconnect to our regular routes through a series of experimentations with different modes, rhythms and conditions of travel.

3. Mapping Our Commutes

Laura's commute is made up of four parts. The "rhythms" of her journey between Sandy Beach, where she lives, and her place of work in Glasgow, incorporate a ten minute drive from her house to the ferry terminal in Dunoon, a 25 minute ferry across the Clyde estuary, a 44 minute train journey from Gourock to Glasgow and a ten minute walk up Hope Street from Central Station to the RCS.

David's 50 minute car journey from Glasgow to Ayr takes him initially through a maze of residential streets, roundabouts and traffic lights to the ramp at the end of the Great Western Road where he joins a short stretch of the M8 south through the city and over the River Clyde to join the M77, which cuts through the Ayrshire countryside before meeting the A77 at Kilmarnock. This is

"Route 77", according to the sign for Balbir's Restaurant which marks the final stage of his journey, and he has always enjoyed the connection to the great American highway.

We started this project by keeping commuting diaries, documenting the range of activities undertaken during our regular journeys to and from work. Laura kept a notebook record of her journeys every day between September and December 2013:

> Tuesday 15th October 2013
>
> 7:00 leave house, 7:15 boat, 7:47 train, 8:45 arrive at work.
>
> 17:25 train, 18:20 boat, 19:00 arrive home.
>
> On both journeys today I read Rebecca Solnit's A Field Guide to Getting Lost. I like it and a statement in capital letters on the back cover reads 'NEVER TO GET LOST IS NOT TO LIVE' [27].
>
> Monday 28th October 2013
>
> 16:55 train, 17:50 boat, 18:30 arrive home.
>
> I take my suitcase on the train and finish reading the book my Aunt Mo has lent me called Olive Kitteridge. It is falling to bits as she read it on her travels to India. It travelled with me to Manchester but was not read [5].

Reflecting on her diary, Laura observed behaviours that she performs repeatedly; mainly reading and sleeping! She enjoyed noticing how many books had been passed on to her from friends and family, and how the books were on their own journeys through "striated space" of commuter routes, being marked by their journeys and the bodies that facilitated their wanderings.

Meanwhile, limited by his role of driver, David used the hands-free mobile phone device in his car to phone spoken diary entries to himself, which he later transferred to MP3 files and transcribed.

> It's drizzly but the sun is cutting through the clouds and picking out the oranges and browns of the autumn trees. And it makes me think this is so much better than having to commute on an underground train in a busy city. I really value this time to drive through the countryside every day [28].

This process of keeping commuting diaries prompted us to look for ways to open up our journeys and to more meaningfully connect with our commutes and the spaces that we move through. Entries immediately reveal an active and productive use of travel time. However, this is too often a movement that drives "through" landscape, rather than travelling "into" it; and while our modes of transport carry their own particular risks, they generally operate through a comfortable dislocation from our wider environment. Less expensive, and ecologically damaging forms of transport—such as walking and cycling—allow a different type of engagement with commuting routes. While the choice to walk and cycle is significantly determined by levels of network capital, it is also the case that walking and cycling have the potential to be more exciting, pleasurable, interesting and relaxing than other modes of transport, and that the physical dimension of these more basic modes has a range of benefits to the regular commuter [29]. This raised an important question for our project: how could we temporarily shift our commuting practices to other forms of travel that might offer us a more corporeal, visceral and hands-on encounter with our routes? And what "benefits" might we accrue through this process?

There are significant parts of our journeys that more closely resemble Sennett's version of desensitised travel [6]:

> Monday 4th November 2013
>
> 6:30 leave house, 6:45 boat, 8:45 arrive at work.
>
> There is a boat refit on from the 4th to the 16th of November so I have to leave my house at 6:30 to get in for 9am. There is a long gap between the boat and the train and the train at this time of the morning is very cold. I read Uncle Silas (on loan from my sister) and doze on the train.
>
> Sunday 24th November 2013
>
> 18:00 Western ferry, arrive home 19:00.
>
> I drive so cannot do anything with the time [5].

There are also moments when we are prevented from undertaking the necessary journeys. In general, Laura's journey is much more prone to disruption than David's due to the modes of transport involved:

> Thursday 5th December 2013
>
> Both boats are off due to high winds and the road is flooded so I cannot travel to work today. This is the first time this has happened this year and I miss the second year show that I am supposed to be assessing which I am really disappointed about [5].

The commute is often perceived, and experienced, as a boring, stressful and frustrating form of mobility [30] (see Figure 2). However, for Edensor this "everyday realm" of "habit, routine, unreflexive forms of common sense, and rituals [...] paradoxically also contains the seeds of resistance and escape from uniformity [through] the intrusions of dreams, involuntary memories, peculiar events, and uncanny sentiments" ([31], pp. 154–55). The etymology of the word "commute" is from the Latin "*commutare*", which means "to change, transform, exchange" [32]. We are interested in the potential of this repetitive journey from home to work as creative or transformative. Braidotti claims: "the imagination is not utopian, but rather transformative and inspirational" and we wanted to use this collaborative project as an opportunity to experience these journeys anew, to revise them and re-imagine them" ([20], p. 14). We have found that the experience of being in transit lends itself to such moments of imaginative departure.

Laura Watts and Glenn Lyons account for the imaginative dimension of travel, arguing that movement leads to an "ambiguity of place", a "liminality" that can foster "a valued sense of creativity, possibility and transition" ([33], p. 109). We set out to disrupt the regular patterns of our commutes and have used the "gift of travel time" to mobilise a critical engagement with our commuting behaviours [9]. This has led us to seek alternative commuting methods, which use different rhythms, physicalities, routes and durations, and opens up our journeys to riskier, messier forms of mobility. Where we usually travel by car, ferry or train, we therefore set out to discover what walking, cycling, swimming and boating might teach us about our journeys to work.

Figure 2. The road to Glasgow.

Four key concepts define the theoretical landscape through which we travelled. However, as our experiences have shown, there are no clear boundaries here. We are moving through heterogeneous and multiple spaces, and routes interweave, blur into each other and amalgamate. These concepts offer us a structure, then, but should not be understood as discrete, fixed categories: first, convergences and divergences of paths and routes and their relationship to the boundaries and barriers that contain and define them; second, notions of becoming, changing and transferring; third, patterns and rhythms of commuting and ways of documenting and analysing these processes; and fourth, the corporeality of commuting as an embodied practice. Together, these concepts offered us an analytical framework for our exploratory journeys, and a set of "guidelines" for the process of recording and representing these journeys, and we paid careful attention to these aspects of our alternative commutes. They provide a loose structure for the map that is beginning to emerge. In the remainder of this article, we present a textual and photographic "deep map" of our walked journeys, followed by individual representations of swimming (Laura), cycling (David), and boating (Laura). This is followed by a brief discussion of the postcard project conducted by Laura on the Gourock to Dunoon ferry route, in order to open up our personal experiments to consider, and learn from, other people's experiences of commuting (see Figure 3). We conclude the article with our suggested "strategies for performing the daily commute" in the hope that others will take up our invitation and map their reimagined regular journeys.

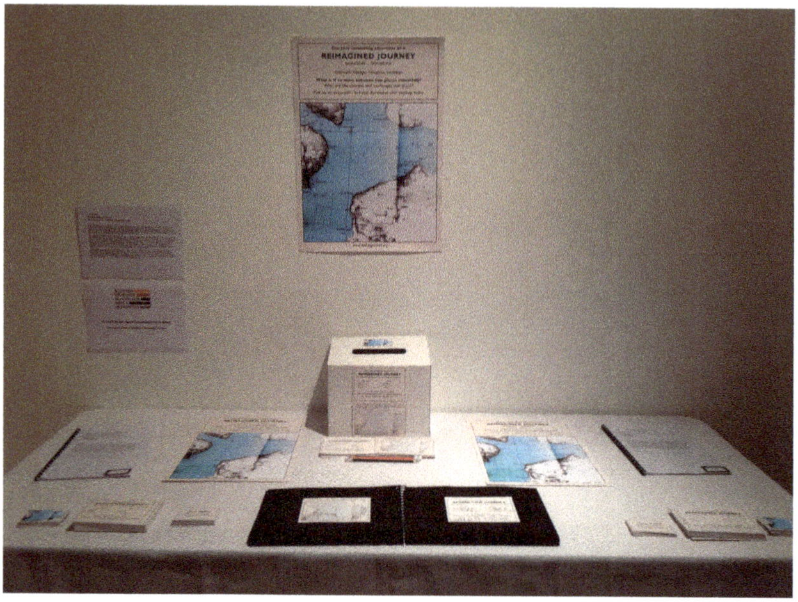

Figure 3. Deep mapping display at the Lighthouse, Glasgow.

4. Walking (Laura Bissell (LB) and David Overend (DO))

For Braidotti, "nomadic subjectivity" is an opportunity "to identify lines of flight, that is to say, a creative alternative space of becoming that would fall not between the mobile/immobile, but within both of these categories" ([20], p. 7). This linking of "lines of flight" with "a creative, alternative space of becoming" offered us a valuable starting point and in March 2014, we both set out to walk our commutes. David walked 48 miles in three days from his home in Glasgow to the University of the West of Scotland in Ayr. A few days later, after a ferry journey from her home on the Cowal Peninsula, Laura walked 27 miles to the Royal Conservatoire of Scotland in Glasgow over the course of one day. Ostensibly, our commutes close down lines of flight to fixed paths across "striated" landscapes [19]. These walks aspired to the opposite—open routes through smooth spaces. What follows is an account of the beginning of our research journeys made up partially from our written responses and critical reflections on our reimagined commutes.

Shortly after leaving his flat, David followed a path along the River Kelvin. In his written account of the walk, he reflects on the way in which our routes through cities often follow fixed paths:

> I [...] realise how much of the city is cut up and divided by fixed routes and paths. This grid of different boundaries and trajectories comprises bridges, footpaths, cycle lanes, roads, railways, canals and flight paths [...] But as I walk from point to point, following some routes and crossing others, I do not feel hemmed in or constrained. There is a freedom in this journey and an exhilarating sense of moving beyond the prescribed uses of urban space [34].

For both of us, undertaking our journeys independently, we enjoyed this sensation of transgression or resistance. As Deirdre Heddon points out, walking allows a potentially "resistant" engagement with landscape—resistance "to habit, to capitalism, to rules, to expectations" (and, suggests Smith, "to fashion") ([18], p. 198); [35], p. 104). Our diaries reveal numerous occasions when we moved against the habits, rules and expectations of our commutes (see Figure 4). We also observed the "resistant" behaviour of other commuters. For example, Laura recounts a moment when a fellow pedestrian breaks out of a designated route:

> as I am about to walk around the cordoned section to allow me to move towards Gourock, the woman in front of me undoes the catch on one of the metal gates, opens it and walks through [...] I follow her lead, glad of this minor transgression of the authoritative paths and designated walkways that the area by the water assigns [36].

(a) (b)

Figure 4. Fences mark the boundaries of prohibited areas on both walked routes.

We are wary about the claims that can be made for such activities and want to avoid what Doreen Massey dismisses as "the least politically convincing of situationist capers" ([37], p. 46). However, these moments of quotidian transgression offered many important formative experiences during our walks, many of which involved a rupture or break in an established pattern or rhythm that we have previously felt frustratingly locked into. Employing Braidotti's "myth" of the nomadic subject to "[blur] boundaries without burning bridges" ([20], p. 26), these moments of transgression during our walks allowed us to reimagine our commutes as something other than functional and transitory journeys through the spaces between home and work.

Braidotti's focus on process and "becoming nomad" is analogous to looking for a marker on a road or path: "the nomadic subject is not a utopian concept, but more like a road sign" ([20], p. 14). For much of Laura's early route, particularly in the Argyll and Inverclyde stages, the road signs are supplemented by another with the Gaelic version of the town name. Often there are four signs which read as a list: district, town, PLEASE DRIVE CAREFULLY, and the Gaelic version (see Figure 5). Laura notices that a large number of the signs also have stickers on them, added by people. Signs are objects that invite us to look at them for information or guidance, but these have been augmented by the public, creating an alternative meaning to that which was intended. What the signs are signifying is also in a state of becoming; the meaning malleable and open to interpretation, obliteration, and subversion.

Figure 5. Road signs for Greenock on Laura's walked route.

These signs served as regular route-markers and they established a rhythm to these walks—marking progress by miles walked and hours passed. However, because "all spaces are dynamic and continually pulse with a multitude of co-existing rhythms and flows", some rhythms became more apparent than others ([12], p. 200). During his walk, David reflected on the rhythms of his regular journey by car:

> In my car, the rhythm of the road takes precedence as my 'insulated mobile body' remains oblivious to everything else ([12], p. 200), but this walk allows me to seek out, sense and immerse myself in multiple rhythms including weather, seasons, animals and people who use and inhabit the landscape that the road cuts through. As I approach Symington, there is a brief moment when I can see the A77, the Firth of Clyde and a Ryanair plane taking off from Prestwick—rhythms and flows coexisting [34].

Earlier in his walk, crossing Eaglesham Moor, the rhythms of Whitelee Windfarm create a unique relationship between technology and landscape (see Figure 6):

> We surrender to the hypnotic quality of the turbines, which offer a regularity and rhythm approximating the effect of music. Our ambulatory rhythms sync up with the rotations of the blades and we adjust our pace accordingly. One foot after the other meets one rotation after the other. At the same time, in this exposed location, rain beats into our bodies and blasts our faces with icy water. The wind is blowing in tremendous gusts, which drown out all other noises, but in the occasional dip, our ears tune into another constant sound—the faint, humming drone of the windfarm [34].

Figure 6. The turbines of Whitelee Windfarm.

Edensor points out that commuting takes place within "routinised, synchronic rhythms [which] are bureaucratically regulated and collectively produced" ([12], p. 196). In the systematised rhythms of motorways and shipping lanes, these natural rhythms—of weather seasons, animals and people—can easily be obscured. For Henri Lefebvre [38], the establishment and regulation of rhythms is one of the key ways by which late capitalism produces and controls space. Similarly, Deleuze and Guattari identify "the function of the sedentary road, which is to *parcel out a closed space to people*, assigning each person a share and regulating the communication between shares" ([19], p. 380). By attempting to establish a "nomadic trajectory" we aspire towards the opposite: movement within what Deleuze and Guattari refer to as "*an open space*" ([19], p. 380) and the potential to take up Edensor's project of "tailor[ing our] journeys in accordance with [our] own strategies, imperatives and feelings" ([12], p. 196). However, this was not always as easy as we hoped. As David reached the southern boundary of the golf course at Pollock Park, his route was blocked by a high metal fence:

> My dilemma is whether to add more miles to my journey by trailing the perimeter looking for an exit, or to attempt to cross into the adjoining field. The solution presents itself as I notice a missing railing which I suspect has been removed deliberately to open up a

> walking route. I make my way through woodland and clamber over another lower fence and land in squelchy mud. Following the high wire fencing that separates the field from the M77, I enjoy a feeling of subversion and recall Braidotti's vision of nomadism as 'the intense desire to go trespassing' ([20], p. 66). Whether or not this is strictly trespassing, it is a moment that embodies the spirit of this journey, moving against prescribed uses of space and reimagining my route as a space of creativity and transgression. However, just as I cast myself as the heroic psychogeographer, I come face to face with a large highland cow. Its menacing stare, sharp horns and slow, deliberate movement towards me make me nervous and I look around for an exit route in case things turn nasty. Unfortunately, I am now too far from my entry point to retreat, and the section of fence that I am beside is too high to easily climb over. The cow lunges forward and a surge of adrenaline catapults me over the fence before I have time to think. On my way over, I scratch my shin and slightly cut my finger. Writing up my notes almost exactly a week later, the cut is still faintly visible - a corporeal document of this encounter [34].

These moments of pushing, testing or breaking the body punctuated our walks, which were defined by physical endurance as much as situationist escapism. Opening up the possibility of corporeal intrusion into our commutes provided an alternative to Sennett's passive commuter bodies [6]. Over the following months, we took this part of our investigations a step further as David cycled to work, and Laura considered ways to negotiate the Clyde Estuary.

5. Swimming (LB)

The 27-mile walk I did in March 2014 had been a very enjoyable experience: the day was sunny and warm and I took great pleasure in noticing elements of my journey that the usual pace does not allow. "Walking is pedestrian" as Smith says and despite the blisters at the end of the longest day of walking I have ever done, I was aware that it was just *walking* ([39], p. 14). Most people can journey by foot and walk from A to B and I had not undertaken any specific training for my epic day of walking from dawn till dusk. The sea-swim, however, was a different story and the sense of anticipation and fear around undertaking this physical challenge seemed to grow as my various attempts at swimming from Gourock to Dunoon (across the Clyde Estuary where my ferry crosses) were usurped. Initially I had hoped to complete the swim prior to David's and my visit to Lapland in April for a mobile train conference [40], however freezing temperatures and rough seas at this time of year made this impossible. I intended to swim the Clyde on the 28 June as part of an organised sea-swim and had ventured out in my wetsuit for some trial swims with my dad rowing alongside me in his dinghy in case I encountered difficulty. After training in the local pool, my first sea-swim was a pleasant surprise. The water was a reasonable temperature, there was nothing awful floating by and I tried very hard not to think of what might be happening in the deep black murk below. One day I even had a friendly seal swimming alongside me and the sense of pervading dread I had felt since I decided to do the swim began to recede. Every day since I made the decision to swim the ferry journey I had looked slightly differently at the body of water I crossed on the boat each morning and evening, assessing the waves, the tides, and what I might be sharing the water with. One week before the event I ended up in hospital with a soft tissue infection in my leg and made the decision not to do the organised swim. Disappointed and immobile I asked my dad to accompany me later in the summer in his boat so I could complete the task I had set myself.

As the warm summer weather dwindled there was one weekend left at the end of August which would be suitable and the 23 August was decided as the date for my next (and final) attempt. The morning was calm and I travelled to my parent's house at 8:00 a.m. to set off in my dad's boat from the West Bay to the Cloch lighthouse on the other side of the estuary. I think my family had hoped that by the end of the summer my will to do the swim would have waned and my dad had seemed anxious about the journey in the days prior to it. When we launched from the West Bay to motor the boat across, almost immediately we lost a small part of the outboard motor in the water. I stripped to my wetsuit and dived into the freezing sea to try to retrieve it but to no avail. This sobering start meant that we had to try to find another part for the motor but after several attempts we realised that it was not going to work and my dad began the process of rowing us across the sea. What I did not realise at the time was that in one of the attempts to start the outboard motor my dad had staved his middle finger but he did not tell me this as we embarked upon our slow transition across the estuary, nor would he let me row so I did not exert myself before the swim. Due to our slow and calamitous start we did not reach the far side and the Cloch lighthouse where I had intended to start but with this in near sight, I lowered my already soggy wet-suited body back into the water to begin the swim back to land. I filmed part of this swim using a camera attached to my head, but, like David's experience of filming when cycling (see below), the film is skewed, a sideways shot of the blue sky peppered with fluffy clouds with a triangular sliver of sea sloping across one side of the screen. The occasional shot of my dad in his dinghy comes into view as I turn to speak to him as we assure each other we are alright. Although the swim was tiring and I subsequently caught a chill (probably from having to dive into the water sooner than intended and then sit in a damp wetsuit while we rowed out) I remember thinking as I was doing it how much I love being in the sea. The buoyancy of your body that the salt water provides makes for an exhilarating feeling of floating. At times I felt as though I had to really push my fins into the sea in order to propel myself forward as they felt as though they were continually floating upwards towards the blue sky as though seeking air. Back at shore, the relief was palpable, no one had died in the attempt (as my sister had feared) and although we did not make it from point A to point B as intended, I felt as though I had swam enough of the journey to justify having done it. Afterwards, I thought about my dad. I had not realised how stressful my attempt at doing this would be for him. Watching the video back I think about his finger and how he did not tell me about it because he did not want to disappoint me in not being able to complete the journey I was so eager to do. I also reflected on how rarely in these re-imaginings have the start and end points (and journeys in between) been *exactly* the same as the commute that I do every day. Perhaps a re-imagining is also a re-inventing, a re-creating of these journeys and their purposes and meaning (see Figure 7).

Figure 7. Laura in Dunoon after her seaswim.

6. Cycling (DO)

According to the old maxim, "it's just like riding a bike", the skill of cycling, once learned, is never forgotten. As I had not cycled for many years, I reassured myself with this thought as I set off on 11 May 2014 from Glasgow to Ayr on a borrowed old mountain bike with a Go-Pro camera fixed to the handle bar. As it turns out, cycling is not at all like riding a bike. I nervously manoeuvred my way through Finnieston and across the River Clyde, brakes squealing and wheels wobbling. At some point in the first few minutes, unnoticed by me, a jolt knocked the camera out of position and for most of the rest of the journey only my anxious face can be seen, framed by the overcast sky (see Figure 8).

Figure 8. David during his commute by bicycle.

Before long, the bike and I fell into our stride. I corrected the camera angle, breathed in the cold spring air and propelled myself forward into the route. Sometime later, just north of Moscow (yes, there is an Ayrshire village called Moscow!) hail stones the size of grapes began to shoot from the sky as the weather turned savage—reddening my exposed legs and soaking me in freezing rainwater. I searched in vain for a shelter and eventually found an old barn where I sheltered for a few minutes. As I waited I realised that the muddy ground was swarming with insects crawling out of the deluge into the safety of the manmade structure. I weighed one discomfort against the other and decided to plough on through the downpour (I could hardly get any wetter).

Cycling occupies the middle ground between walking and driving: fast enough to feel a tangible sense of progress at any given moment; but slow enough to feel a corporeal connection to the landscape. I remember this journey vividly a year later. The exhilaration of the wind in my face as I sped downhill along country lanes, the mud splattering all over me, and the chilling soaking as I approached Ayr. As an extremely irregular cyclist, I do not have a sense of cycling as a mode of commuting, which can be perceived as just as "forgettable, ordinary, banal" as my commute by car ([41], p. 291). This experience, for me, was like Phil Ian Jones' artistic experiments in route-marking by bike, and allowed me to engage with this form of mobility as "sweaty, visceral, emotionally intense and highly embodied" ([41], p. 291). When I arrived at my hotel later that afternoon, it was one of the most warming and relaxing moments of arrival that I have experienced. This is a welcome return from the *détournement* of the journey, and it reminds me of Smith's observation that "the permanent drift disappears the drifter" ([18], p. 139). Returns, homecomings and finishing lines. Unlike nomadic journeys, all these exploratory commutes have an intended endpoint.

7. Boating (LB)

In the Victorian era when Glasgow was the "second city" and the tobacco merchants were making their fortunes, the local seaside town of Dunoon grew as the wealthy merchants built holiday homes within sailing distance of the city. Dunoon became a popular holiday resort and steam-powered boats transported thousands of holiday makers "doon the watter" of the river Clyde into the Firth of Clyde for summer holidays and Glasgow fair weekend. This continued until the 1960s when a controversial American naval base was built in the nearby Holy Loch and the rise of cheap package holidays to warmer climes meant that the popularity of Dunoon as a holiday destination declined. My grandparents met at in Dunoon on a Glasgow fair weekend and my father spent a lot of his childhood holidaying in Dunoon and subsequently my sister and I spent many rainy summer holidays following suit. My parents moved to Dunoon six years ago and I moved to the nearby village of Innellan the year after. The faded elegance of Dunoon as a popular Victorian resort and the history of the town includes a shift from sea travel to alternative means. The Waverley is the world's last sea-going paddle steamer and epitomises a mode of transport that was once the norm but is now all but obsolete. Almost my entire commute can be done over water and so I wanted to travel by boat from Glasgow through the narrow banks of the Clyde river in the city centre as it opens out into the Clyde Estuary and out to sea.

I undertook my usual commute by car, ferry and train to get to Glasgow on the morning of 22 August, one of the Waverley's last sailings of the summer season. Boarding the boat at Glasgow's Science Centre (the recent addition of new bridges over the city centre stretch of the Clyde makes this the closest point that larger vessels can travel to and from) the boat is busy with day-trippers heading to various seaside resorts (the steamer will stop at Kilcreggan, Dunoon, and Rothesay). There is something very alluring about the Waverley's design and over all three levels of the steamer there are details of design and engineering that are notable. In the midsection of the boat is the engine room where the mechanics and workings of the engine are on display. The large paddle can be seen moving through the water through the portholes at each side of the boat.

The journey "doon the watter" is very pleasant. I stay on the port side of the boat so I can film the commute I do every day from the perspective of the water (see Figure 9). I am pleased to notice familiar landmarks from my walk and the shift from the scrapyards and building yards of the city into the yellow fields of Inchinnan is beautiful. Once almost at Greenock I move about the boat and film various parts of my journey around the steamer. We stop at Kilcreggan, a small village near Helensburgh, before journeying on to Dunoon. I think how lovely the pier looks as we approach the town and also think about how on my daily journey I do not look at this anymore, or do not see it in the way I once did. The experience of being on a "daytrip" on the Waverley allows me to see Dunoon as a visitor might on arrival. The antiquated mode of journeying has also made me think about Dunoon's transition from holiday town to declining working town and I wonder if the tourists who visit Dunoon on the Waverley transport the destination back to a former identity—a seaside town visited for pleasure during the summer holidays as I experience it on this day. It was this experience that encouraged my "everyday commuting excursions" postcard project with designer Rachel O'Neill, where we wanted to use some of the nostalgic aesthetic that often appears in connection with the golden age of steam travel to ask contemporary boat travellers to reimagine and map their journeys and to see the spaces between home and work a-new.

Figure 9. The Waverley.

8. Everyday Commuting Excursions

One of our mapping experiments, which Laura developed with O'Neill, was a set of posters and postcards encouraging commuters on the Gourock to Dunoon ferry to reimagine their everyday journeys as "excursions". When Laura traveled down the Firth of Clyde on the Waverly Steamer in August 2014, she acknowledged that the arrival at Dunoon ferry terminal by steamer as opposed to ferry made her see the Victorian pier and the town beyond differently. By traveling on a historic mode of transport the feeling of being on an "excursion" as a "daytripper" rather than a commuter encouraged a "seeing a-new" of the familiar site. Inspired by this and by the nostalgic posters portraying the golden age of steam travel on the Clyde situated within the newly developed Gourock train station, Laura wanted to engage other commuters in a creative act of mapping that would encourage them to reimagine their everyday journey by boat and to "see a-new" the space between home and work. The design of the materials was influenced by ferry brochures and posters from the 1950s and commuters were asked to offer a creative response to a map showing the area travelled over water between the two ferry terminals. By obscuring the landmass and emphasising the sea using colour and detailed markings, the design was intended to ask people to consider what this space and time travelling over water could be. Participants mapped, drew or wrote in the space provided on the postcard to offer an alternative mapping of the spaces between home and work and posted it into a postbox on the Argyll Flyer ferry (see Figure 10). These responses were collated into a book which was displayed at the Lighthouse in Glasgow in July and the maps completed by members of the public can now be found on the Making Routes website [2].

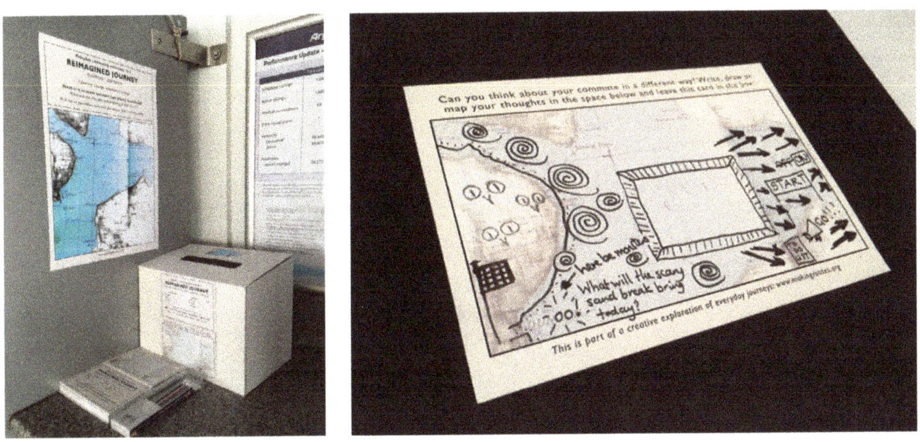

(a) (b)

Figure 10. (**a**) Postcards at the Gourock ferry terminal; (**b**) One of the public responses.

The responses of the public were varied with some participants drawing feelings they had about their journey and others offering practical solutions for the "problem" of the journey. One postcard depicted an arm with a floating balloon attached over Dunoon and a ball and chain on an ankle for Gourock implying a feeling of lightness and freedom when "home" and a heaviness and responsibility connected with the "work" side of the map. Some of the contributions conveyed the history of the area with one showing a "map of memories of Clyde piers" with sketches of now defunct or absent piers as boat travel has become less popular. Others used the postcard to convey the social aspect of the journey with one participant writing: "The boat is a place to meet friends old and new" and another explaining it was a time to "exchange knowledge of local community events". Some simply drew crosses on the map to depict the journey and one childlike drawing of a boat on the map was also offered. Practical solutions to some of the problems of travel also featured with some offering suggestions of a walkway between the two points and the simple statement "UNDER 18 GO FREE", which hints at some of the issues young people face in Dunoon with many of the colleges situated over the water in Greenock. By asking commuters on the Dunoon to Gourock ferry to reimagine their journey as we had been doing with our respective commutes we were able to share our ideas with a specific commuting community. Encouraging people to share their "maps" (both literal and imaginative) also gave us an insight into how our strategies or prompts might also offer us new ways of thinking about our routes.

9. Strategies for Performance

As we attempted to reimagine our commutes, the overarching aims of the journeys and routes remained consistent but the pace, rhythm and embodied experience and exertion varied greatly depending on the mode of travel or transport. The answers to the questions "where have you come from?" and "where are you going?" were always the same, and always interchangeable (either "home" and/or "work") for every journey we undertook. Within this "formula" for these journeys we began to think of our experiments as strategies or variables for this predictable and repetitive commuting route. By varying the method by which we filled the space and time between "home" and "work" we were able to challenge our own understanding of these quotidian and functional journeys. Our walking, swimming and boating journeys introduced a range of approaches to mobility that allowed us to depart from our established rhythms and routes and reconsider the liminal transitory space between the fixed sites of "home" and "work". The start and end points remained (roughly) the same, however the journeys themselves offered a space of possibility—of becoming—that differed for each reimagining. Furthermore, what started out as a relatively discursive and academic exploration of commuting—an attempt to apply a nomadic thought experiment to our regular journeys—soon became far more physically and emotionally affective than we had originally considered. Our deep map contains accounts of venturing out into woods, estuaries and country lanes, often inexperienced and underprepared for a range of hostile and challenging environments and modes of transport. This introduced unpredictability, risk and endurance as key elements of our journeys. This aspect of our experiments in commuting will benefit from further research, and opens up potential routes for ongoing work in this area. In this final section, we distil some of these experiences into a list of "strategies", which are offered for others to take up our project of "performing the daily commute". If these strategies lead to a greater degree of uncertainty, and lead those who undertake them into challenging situations, this is all for the better. However, be careful out there.

(1) Alter the pace of your commute to take from dawn until dusk, or for a full weekend, week, month or year
(2) Perform your commute in the hours of darkness
(3) Undertake your commute on an obsolete/historic mode of transport

(4) Spend the night at your place of work and commute in the opposite direction as you travel home the following morning
(5) Take your friends and family with you
(6) Create a guided tour for your route
(7) Leave extra time for unplanned detours and unexpected encounters
(8) Plan an alternative route and take this every day for a week
(9) Research the history of your route and re-enact previous travellers' experiences
(10) Record every human exchange along the way—every nod, wave and conversation

For those who are tempted to employ these strategies, or to invent new ones, we encourage the sharing of experiences and images on the Making Routes website [2]. As this project continues to develop, this site will extend this mapping project to maintain the deep map as an open, dynamic and multi-layered document, rather than fixing any "findings" in this journal article. Nonetheless, our process so far allows us to draw some initial conclusions.

10. Conclusions

On the first day of his walk, David cheated and briefly broke the task of his continuous journey on foot, returning by taxi to his home rather than spending the night in a soulless budget hotel. The following morning, returning as a car passenger to resume the walk where he left off, he noted that he was "already thinking of [his] commute differently, pointing out places he walked through the day before and noticing features of the route of which [he] had previously been unaware" [34]. Similarly, Laura noted her aspiration that the walk would allow her to "re-experience my commute in an active and embodied way" in order to "emancipate" herself from what Braidotti refers to as "the inertia of everyday routines" [36]. Our experiments in commuting have shown potential in this respect as both of us feel that applying our various performative experiments to our commutes, has changed the way we experience our daily journeys in several important ways. The postcard project "everyday commuting excursions" also allowed us to encourage commuters on the Dunoon to Gourock ferry to creatively map their thoughts and to reflect on these routine journeys and what the space between home and work means to them.

First, although we continue to follow the same roads, railways and ferry routes, we no longer feel as limited by, or contained within, these prescribed patterns. We have developed the appetite and tools to go "off route". In this sense, we have inserted "nomadic shifts" into our regular journeys [20]. While the sites of "home" and "work" remain fixed, by applying the first strategy to "alter the pace of your commute to take from dawn until dusk, or for a full weekend, week, month or year" the time and space in between these places has been opened up to the possibility of a deeper engagement with the landscapes we previously (and continually) traversed at great speed.

Second, Braidotti's insights into nomadism as a way of reimagining subjectivity have allowed us to consider the commute in terms of "becoming" [20]. The change in time and space as we travel these repetitive routes provides a liminal space for commuters to be always changing through new encounters and experiments, even when regularity and repetition are the dominant modes of engagement with a particular route. The sense of departing, moving, traveling and arriving all indicate moving in, through and out of places and the transitory and fluid sense of fixed time and space allow for a space of creative imagining. Strategy three: "undertake your commute on an obsolete/historic mode of transport" can assist in this exploration of time and space, as can strategy nine: "research the history of your route and re-enact previous travellers' experiences". All of the "findings" from these exploratory journeys deepen the historical and geographical range of the map.

Third, we have encountered multiple rhythms that are co-present with our commutes, whether synchronously or asynchronously. Our walking, cycling, swimming and boating journeys have offered us an opportunity to encounter this rhythmic multiplicity beyond those elements that are foregrounded during our regular commutes. The physical exertions of our alternative commutes have allowed for an understanding of the rhythms of the body in relation to the landscape we have moved through depending on how we travelled. We have attempted to capture this rhythmic variety through our mapping process, moving from anecdotal to academic registers and shifting our mode of documentation as we travel.

Fourth, we have engaged in our commutes in a more embodied way and pushed ourselves beyond the ostensibly passive bodily experience of commuting. Smith says "the walk can bring you and your body into new connections through its aches, blisters, shivering and sweating, dehydration in intense heat, dizziness, pain, exhaustion, alienation" ([39], p. 62). David's exhaustion and painful legs from cycling allowed him a new appreciation of rest and shelter, while Laura's fear of whether she would physically be able to complete the sea swim and this challenge moved her beyond her perceived bodily capabilities. These journeys have helped us develop an understanding of a more overtly corporeal engagement with the space of our commutes. We had the cuts, bruises and blisters to prove it—a form of deep map temporarily etched into our bodies.

As we continue our research in this area, we will find new ways to reimagine and map our regular journeys, as we aspire to become (more) nomadic commuters. We are going to keep journeying together and hope to explore the potential of the repetition of our journeys over a much longer period of time. What is it for a body to move between two places repeatedly? What is the "change" or "becoming" within the body through this process? By deep mapping our commutes we have learned more about the places we move through, our personal narratives and perspectives on our travels and how our experience and memory provides a sense of seeing anew the journeys we do every day. The idea of "mapping" has been literally explored through Laura's experiments on the ferry where she asked people to "map" their journey while undertaking the boat journey from Dunoon to Gourock, but the framework of "deep mapping" has allowed us to conceptually map our journeys through a series of embodied alternative experiences opening up the possibility of transgression and reimagining. The challenge for us now is to keep the map open and evolving in order to maintain a sense of our commutes as spaces of becoming, and to continue to reimagine what our journeys are and the potential of what they could be.

Author Contributions

These authors contributed equally to this work.

Conflicts of Interest

The authors declare no conflict of interest.

References

1. Laura Bissell, and David Overend. "Rhythmic Routes: Developing a Nomadic Physical Practice for the Daily Commute [keynote paper]." *Scottish Journal of Performance* 2 (2014): 29–54.
2. Making Routes. Available online: www.makingroutes.org (accessed on 5 April 2015).
3. The Arches. "Making Routes Launch Afternoon." 2011. Available online: www.thearches.co.uk/events/arts/arches-live-2011-the-arches-and-the-university-of-the-west-of-scotland (accessed on 2 February 2015).
4. David Overend. "David Overend is Misguided in Ayr." 2012. Available online: makingroutes.ning.com/profiles/blogs/david-overend-is-misguided-in-ayr (accessed on 2 July 2014).
5. Laura Bissell. "Commuting Diary." Unpublished manuscript, 2013–2014.
6. Richard Sennett. *Flesh and Stone: The Body and the City in Western Civilization*. London: W. W. Norton, 1996.
7. Jean Baudrillard. *The System of Objects*. London: Verso Books, 2005.
8. Michel de Certeau. *The Practice of Everyday Life*. London: University of California Press, 1998.
9. Juliet Jain, and Glenn Lyons. "The Gift of Travel Time." *The Journal of Transport Geography* 16 (2008): 81–89.
10. Patricia L. Mokhtarian. "Travel as a Desired End, Not Just a Means." *Transportation Research Part A* 39 (2005): 93–96.
11. Anthony Elliott, and John Urry. *Mobile Lives*. Oxon: Routledge, 2010.
12. Tim Edensor. "Commuter: Mobility rhythm and commuting." In *Geographies of Mobilities: Practices, Spaces, Subjects*. Surrey: Ashgate, 2011, pp. 189–204.
13. Mike Pearson. *"In Comes I": Performance, Memory and Landscape*. Exeter: University of Exeter Press, 2006.
14. Mike Pearson, and Michael Shanks. *Theatre/Archaeology*. London and New York: Routledge, 2001.
15. Fiona Wilkie. "Site-Specific Performance and the Mobility Turn." *Contemporary Theatre Review* 22 (2012): 203–12.
16. Fiona Wilkie. *Performance, Transport and Mobility: Making Passage*. Basingstoke: Palgrave Macmillan, 2015.
17. Carl Lavery. "Teaching Performance Studies: 25 instructions for performance in cities." *Studies in Theatre and Performance* 25 (2005): 234–36.
18. Phil Smith. *Mythogeograhy*. Axminster: Triarchy, 2010.
19. Gilles Deleuze, and Félix Guattari. *A Thousand Plateaus*. London: The Athlone Press, 1998.
20. Rosi Braidotti. *Nomadic Subjects: Embodiment and Sexual Difference in Contemporary Feminist Theory*, 2nd ed. New York: Columbia University Press, 2011.
21. Mike Pearson. *Site-Specific Performance*. Basingstoke: Palgrave Macmillan, 2010.
22. Tim Cresswell. *On the Move*. London: Routledge, 2006.

23. Tim Cresswell. "Imagining the nomad: Mobility and the postmodern primitive". In *Space & Scoial Theory: Interpreting Modernity and Postmodernity*. Edited by Georges Benko and Ulf Strohmayer. Oxford: Blackwell, 1997, pp. 360–82.
24. Gil Viry, Vincent Kaufmann, and Eric Widmer. "Social Integration Faced with Commuting: More Widespead and Less Dense Support Networks." In *Mobilities and Inequality*. Edited by Timo Ohnmacht, Hanja Maksim and Max Bergman. Farnham: Ashgate Publishing, 2009, pp. 121–44.
25. John Urry. *Mobilities*. Cambridge: Polity, 2007.
26. Janet Wolff. "On the Road Again: Metaphors of travel in cultural criticism." *Cultural Studies* 7 (1992): 224–39.
27. Rebecca Solnit. *A Field Guide to Getting Lost*. Edinburgh: Cannongate Books, 2006.
28. David Overend. "Commuting Diary." Unpublished manuscript, 2013–2014.
29. Birgitta Gatersleben, and David Uzzell. "Affective Appraisals of the Daily Commute." *Environment and Behaviour* 39 (2007): 416–31.
30. Eleanor Mann, and Charles Abraham. "The Role of Affect in UK Commuters' Travel Mode Choices: an interpretative phenomenological analysis." *British Journal of Psychology* 97 (2006): 155–76.
31. Tim Edensor. "Defamiliarizing the Mundane Roadscape." *Space and Culture* 6 (2003): pp. 151–168.
32. Oxford Dictionaries. "Commute: Definition of commute." 1999. Available online: http://www.oxforddictionaries.com/definition/english/commute (accessed on 28 March 2014).
33. Laura Watts, and Glenn Lyons. "Travel Remedy Kit: Interventions into train lines and passenger times." In *Mobile Methods*. Edited by Monika Büscher, John Urry and Katian Witchger. Oxon: Routledge, 2011, pp. 104–18.
34. David Overend. "David Overend Walks Route 77." 2014. Available online: makingroutes.ning.com/profiles/blogs/david-overend-walks-route-77 (accessed on 31 March 2014).
35. Deirdre Heddon. *Autobiography and Performance*. Basingstoke: Palgrave Macmillan, 2008.
36. Laura Bissell. "Walking Gourock to Glasgow: Reimagining my Commute." 2014. Available online: makingroutes.ning.com/profiles/blogs/walking-gourock-to-glasgow-reimagining-my-commute (accessed on 11 April 2014).
37. Doreen Massey. *For Space*. London: SAGE, 2005.
38. Henri Lefebvre. *Rhythmanalysis: Space, Time and Everyday Life*. London: Bloomsbury Academic, 2013.
39. Phil Smith. *On Walking*. Devon: Triarchy Press, 2014.
40. Laura Bissell, and David Overend. "Reflections on a Mobile Train Conference from Helsinki to Rovaniemi." *Cultural Geographies in Practice*. Published electronically 19 February 2015. doi: 10.1177/1474474015569979.
41. Phil I. Jones. "Performing Sustainable Transport: An artistic RIDE across the city." *Cultural Geographies in Practice* 21 (2014): 287–92.

The Rhythm of Non-Places: Marooning the Embodied Self in Depthless Space

Les Roberts

Abstract: Taking as its starting point the spatiotemporal rhythms of landscapes of hyper-mobility and transit, this paper explores how the process of "marooning" the self in a radically placeless (and depthless) space—in this instance a motorway traffic island on the M53 in the northwest of England—can inform critical understandings and practices of "deep mapping". Conceived of as an autoethnographic experiment—a performative expression of "islandness" as an embodied spatial praxis—the research on which this paper draws revisits ideas set out in JG Ballard's 1974 novel *Concrete Island*, although, unlike Ballard's island Crusoe (and sans person Friday), the author's residency was restricted to one day and night. The fieldwork, which combines methods of "digital capture" (audio soundscapes, video, stills photography, and GPS tracking), takes the form of a rhythmanalytical mapping of territory that can unequivocally be defined as "negative space". Offering an oblique engagement with debates on "non-places" and spaces of mobility, the paper examines the capacity of non-places/negative spaces to play host to the conditions whereby affects of place and dwelling can be harnessed and performatively transacted. The embodied rhythmicity of non-places is thus interrogated from the vantage point of a constitutive negation of the negation of place. In this vein, the paper offers a reflexive examination of the spatial anthropology of negative space.

Reprinted from *Humanities*. Cite as: Roberts, L. The Rhythm of Non-Places: Marooning the Embodied Self in Depthless Space. *Humanities* **2015**, *4*, 569–599.

1. Marooned

> Marooned in an office block or on a traffic island, we can tyrannise ourselves, test our strengths and weaknesses, perhaps come to terms with aspects of our characters to which we have always closed our eyes ([1], p. viii).

Running across a northbound stretch of the M53 motorway one August afternoon in 2013, a backpack of basic supplies and equipment thumping against my spine, my progress was steered towards territory that could lay claim to few, if any, attributes of what might count as "place". Initially conceived of as a kind of autoethnographic experiment, the aim of this exercise was not to tyrannise myself, Ballardian style, as much as to try and "map" a sense of embodied dwelling (my "being there"): to examine how the "non-place" towards which I was headed might play host to a performative study of the rhythms and cadences of a radically liminal space. Hurtling headlong across the motorway, taking strategic advantage of a brief lull in the traffic flow, my goal was thus to maroon my *self* in a space defined in opposition to those routinely inhabited or imagined: a negative space against which the more mundane geographies of everyday mobility might be cast and measured.

The liminal space in question, like that which forms the setting of Ballard's story, is a motorway traffic island. A "concrete island", to borrow from the title of the 1974 novel, is a reasonably apt descriptor, but only inasmuch as this too is constitutive of a negative frame of reference: an island surrounded by a sea of concrete (and tarmac) rather than one that is made of concrete. Indeed, it was partly the lushness and density of the island flora that first drew my attention to this topographic anomaly when driving past it on my way to and from Liverpool. The island's anomalous status (of which more later) was another factor that piqued my initial curiosity, setting in motion the designs of a Crusoe-esque quest spurred as much by the question as to why such a feature existed at all as by that which probed the island's more experiential or "rhythmanalytical" ([2,3]) dimensions. What might be gleaned, I had wondered, from time spent self-marooned on this motorway island; a day and night cocooned by nothing but the hiss and roar of high-speed traffic? What might a spatial anthropology of a such a radically deterritorialized landscape throw up that could not ascertained by other modes of enquiry? (What does the "being there" element bring to the table?) Given the depthless and socio-historically "empty" fabric of this non-place, to what extent might a project of "deep mapping" make sense of a space that, save perhaps the occasional visitation by motorway maintenance workers, has remained cut off from all human contact?

But then had it? This too formed part of what I had hoped to shed some light on. Might I find evidence of others having "made the crossing" and spent time on the island (for whatever purpose)? Might my being there serve in some way to draw out or tentatively sketch an anthropology of this territory? Or might I end up finding myself in search of a rationale for an excursion destined to elicit little that might be construed as anthropologically or geographically insightful, other than perhaps that which speaks to more generic concerns with, for example, "edgelands", "psychogeography", "liminality", "non/more-than-representational geography", or other such modish frames of reference? Posed more directly: Beyond the more broadly conceived goal of exploring the spatial anthropology of non-places—a reflexive enquiry into what could be described as the *negation of the negation of place*—what, exactly, is the *point* of marooning the embodied self in depthless space?

2. Tuning in, Timing Out

To begin to flesh out a provisional response to this question, the idea of "tuning in" provides a useful and suggestive point of departure. Finding the right frequency is to connect or find resonance with a space of communication defined in terms of modulation and rhythm rather than infrastructural and material form. Viewed thus, for those marooned in "non-places" rather than in transit through them, adjusting—mindfully—to the temporal geographies of motorway space is akin to tuning the dial on a car radio. Rather than confronting the "white noise" of traffic that defines a dissonant or transitory moment of encounter (such as my precarious transit to and from the island), the bestowal of prolonged durational time—the lacuna ([4], p. 19) period spent *in the midst of* flow and movement—provides the motorway dweller with a space by which to ease the embodied self into a habituated sense of rhythmic order, as if slowly beginning to make out the shape and contour of a recognisable tune coming through the airwaves. In autoethnographic terms, as a key methodological objective, "tuning in" thus became an integral part of the wayfinding and habituation process by which I familiarized myself with and made sense of the environment I had made my temporary home. By extension, in cartographic terms, "reading the rhythm" of the space is perhaps the most instructive way of thinking about the project as an exercise in "mapping" ("deep" or otherwise) oriented around the central axis of the embodied and situated self.

In the absence of topographic detail, embodied immersion in the liminal spatialities of a motorway island is therefore as much an enquiry into the phenomenology of time as it is that of space or "non-place". More accurately, it is necessary to look upon the temporal geographies under investigation in terms of an experiential "time out" conceived of as a disruption of the temporal flow that otherwise structures a sense of everyday habitus and being-in-the-world [5]. As Edensor notes, "[t]he usually unreflexive sensual and rhythmic attunement to place and familiar space may be confounded when the body is 'out of place'" ([3], p. 5). Accordingly, it is to what Morris has referred to as the "*metaphorai* of the pause" ([6], p. 41) that our rhythmanalytical attention is principally addressed.[1] Concerned less with arrival and departure in terms of the consummatory passage *through* a space or non-place of transit, the experiential space–time of the island is more redolent of an embodied sense of stasis: a temporal unfolding of spatial restriction where movement is defined *in relation to* the immobility and longueur of the liminal "pause" ([9], pp. 82–83; [10], p. 5). Horizons are scaled down to the immediacy of the moment and the ebb and flow of time that is mobilized in vehicular form. From the axis point of the island Crusoe, cast ashore and stranded from a world in perpetual flux, the north- and southbound rhythms of the motorway inhibit any sense of resolution or consummation other than that by which the island-body is brought into being as a time unto itself.

In his study of memory and oblivion, the anthropologist Marc Augé ([11], pp. 55–84) sets out what he refers to as three "figures of oblivion" that structure the temporal dialectics of remembrance and forgetting. The second of these, *suspense*, refers to the cutting off of the present from the past and future. In ritual terms this approximates to the separation phase of Arnold van Gennep's tripartite "rite of passage", or Victor Turner's notion of "communitas"—an in-between state in which the oblivion of past returns and future beginnings (or of points in time and space already departed as well as those yet to be arrived at) shapes a temporary state of suspended *presentness* [12–14]. The motorway island, like that of the motorway more generally, is a liminal space of oblivion par excellence. Memory is afforded little in the way of traction. Anthropological place, in the Durkheimian sense of an organic social space that is "relational, historical and concerned with identity" ([15], p. 77), has limited application in the case of environments such as these. The mirror-like "surface sheen" (the functional prerequisite of a space given over to pure mobility) prevents the slow absorption of time and memory; it can only be "layered" or "embedded" insofar as the mode of transport itself becomes the site of topographic depth ([3], p. 6; [16–18]).

---

1 The word "metaphor" is derived from the ancient Greek for transfer, or transport. Moreover, in modern-day Greece, vehicles of mass transportation, such as the bus and the train, are called *metaphorai* ([7], p. 115). Viewed thus, the "*metaphorai* of the pause" can be understood as the movements that are *metaphorically* transacted at what are otherwise fixed (immobile) points where trajectories of arrival and departure intersect (such as, in Morris's [6] example, a motel—see Roberts [8]).

A brief caveat at this point would be to acknowledge that, yes, Augé's non-place thesis to the effect that "place" in supermodernity becomes emptied and rendered devoid of history, memory and organic social relations is not without its detractors. As Merriman, for example, has noted, to look upon airports or motorways as "non-places" runs the risk of overlooking "the complex habitations, practices of dwelling, embodied relations, material presences, placings and hybrid subjectivities associated with movement through such spaces" ([19], p. 154; [20]). Whilst such a risk might arguably relate more to the peddling of an absolutist conception of non-places (which Augé himself does not subscribe to, noting that place and non-place are like "opposed polarities: the first is never completely erased, the second never totally completed" ([15], p. 79)), this observation nevertheless draws much needed attention to the complex experiential and embodied dynamics of these spaces and holds in check the otherwise default setting that non-places are, almost by definition, intractably place*less*.

Yet to the extent that we could ascribe a (rather clumsy) measure of "non-place-ness" in our analyses of specific spaces and topographies, there is clearly a distinction that could be drawn between, say, a motorway service station or airport terminal and the kind of space that forms the basis of the present study. Embracing the abstraction, solitude, and liminality of this deterritorialized landscape, my curiosity was driven in no small part by the degree to which the island is an exemplar of a radically placeless space, and by what, as a consequence, this might translate to in terms of a strategically enforced practice of embodied dwelling. "What is the *time* of such a non-place?" is a question that most succinctly draws together the various threads of what I conceived the project to be reaching for in terms of its autoethnographic remit.

3. Time Out (of Place): Off-Road Journeys

If motorways are an example of what Dimendberg ([21], pp. 177–78) refers to as "centrifugal space", where the density of the urban centre is radially dispersed throughout what are increasingly abstract, hyper-mobile and virtual spaces and networks, then the "walking cure" ([21], p. 109) prescribed by the *flâneur* or psychogeographer can be looked upon as a kind of counter force. This need not necessarily be centripetal in its orientation, but does at least serve to anchor affects of place and memory in spaces that are otherwise experienced as abstract or in some way drained of local colour and depth. The tactical interweaving of imaginative and embodied worlds with the concrete materiality of everyday urban form—"tramping asphalted earth in alert reverie", to borrow writer Iain Sinclair's suggestive phrase ([22], p. 4)—thus allows for the layering and accretion of anthropological place, if by this we mean places invested with humanistic attributes, the "impure content..."—"...lived time, everyday time, ...bodies with their opacity and solidity, their warmth, their life and their death" ([23], p. 97)—that processes of spatial rationalisation can prove so adept at flushing out.

By extension, the motorway *flâneur*—as illustrated by Sinclair's [24] epic trek around London's M25 orbital motorway—is someone whose cartographic proclivities are steered towards the deep mapping or "deep topography" [25] of landscapes through, across or above which, ordinarily, the subject is centrifugally slung. Morse notes that "[f]reeways are displaced in that they do not lie earthbound and contiguous to their surroundings so much as float above or below the horizon" ([26], p. 197). A topographic study premised on the notion of deep mapping is thus one that is bent on reversing or going against the grain of the motorway conceived of as a rational and functional space. A reworking of Morse's observations along these lines would instead stress that: motorways are *emplaced* inasmuch as they are *made* earthbound and contiguous to their surroundings by the *embodied practices* of a deep cartographer for whom horizons are experienced in the simultaneity of the *here and now*. Presentness—a dalliance with *off-road* space (the chronotope of the road traded for that of the route or way)—denotes *earthiness* in that time is re-routed and made consonant with the rhythms and textures of the ambulant body.

If "time out" represents a temporal disruption of the functional space–time linearity of the motorway, then "off-road" is its spatial correlate (*cf.* [8,27]). Off-road here can refer to the physical edgelands of motorways (such as, in my case, a motorway island situated between carriageways, or those outlying liminal landscapes that hug the orbital colossus of Sinclair's M25). It can also point to the flights of imagination and creative reverie that the mundane features of motorways can help foster [26,28,29]. A phenomenon that will be all too familiar to motorists who regularly drive on motorways is the experience of slipping into "autopilot" for a period of miles/minutes/hours and of subsequently "coming to" with little recollection or awareness of having been driving, almost as if s/he had temporarily been spirited away to another space entirely. In Morse's "An Ontology of Everyday Distraction", she examines the ways that "non spaces" such as freeways/motorways or shopping malls can promote a state of mind whereby the individual is "*disengaged* from the paramount orientation to reality" and experiences a "'spacing out' or partial absence of mind" ([26], pp. 200, 194, emphasis in original). It is instructive to look upon these lapses and moments of "spacing out" not as ellipses (akin to film or television footage that has been "edited out") but rather as negatives through which we encounter the flux of thought and mental *flânerie*: imaginative worlds that project a different inflection on what we might understand by the concept of non-place/space. This is an idea that finds resonance with Osborne's view that non-places may potentially be configured "less as 'empty' or 'solitary' versions of traditional places and more as radically *new ontological types of place*, constituted qua places through their relations to another spatiality" ([30], p. 189, emphasis in original). It also brings to mind Marin's discussion of the "spatial play" of *utopic* practice: utopia (a conflation of the Greek words *outopia* and *eutopia*—"no place" and "good place") conceived of as the "reverse image of this world, its photographic negative" ([31], p. 242). The non-places of the motorway, in other words, refer to both the anthropologically inert or "empty" spatialities of the built environment and the "other spaces/places" of the imagination that the mundane roadscape [29] helps conjure into being.

In director Laurent Cantet's 2002 film *L'Emploi du temps* (the Anglophone title of which is *Time Out*) we are confronted with a protagonist for whom the motorway offers the possibility of escape from a psycho-social world he is increasingly anxious to forget (or "space out"). Vincent is a French middle-class executive who conceals from his family and friends the fact that he has been sacked from his consultancy job. Maintaining the illusion that he has secured a prestigious new position in Geneva working for the United Nations, he spends most of the week in his car driving through the French and Swiss countryside, telephoning his wife from car parks or motorway service stations, and pretending that he has just come out of business meetings.[2] What is immediately striking about *Time Out*'s cinematic geographies [33,34] is the inert quality of the temporal and spatial landscapes that dominate Vincent's everyday world. Time is as dead and empty as the non-places he inhabits. In the absence of any recognisable social framework, "normality" is rehearsed at scheduled time slots (early morning, lunchtime and evening) and in allotted "social" arenas, such as motorway dining areas (and, later in the film, hotel lobbies and corporate reception areas). In these structured moments of contractual sociality Vincent's double life acquires its fullest expression, most notably in the phone conversations he has with his wife, but also in the momentary encounters with other transitory passengers, whose shared anonymity allow him to temporarily maintain the illusion of a social role and identity. In all other respects, movement "fills in the gaps" of distraction. The abstract spaces of the motorway become a welcome refuge from the responsibility and complexity of his situation. Indeed, as Vincent later confides, it was his increasing reluctance to leave his car that precipitated his eventual dismissal from his consultancy position. Driving hundreds of kilometres to attend pointless meetings was, he confesses, the only thing about the job he enjoyed: "Alone in the car…thinking about nothing…smoking and listening to music…"

Time Out offers an immersive illustration of the capacity of motorway non-places to effect a "spacing out" of some of the more mundane facets of everyday life and to render possible the creation of "other worlds": spaces of the imagination into which the self can take temporary flight. In Vincent's case the rhythms of the day are not dissimilar to those that define his normal social routine; the difference lies in the degree to which this normality may be held in abeyance and a creatively re-ordered reality substituted in its place. As a consequence, the constitutive liminality that governs the ritual and experiential potentiality of these spaces is harvested for its essentially *transformative* properties, however short-lived or psychologically precarious these may be. To these ends, it is apparent that another objective of my island sojourn has been to enquire into the changes potentially wrought by an intermeshing of the embodied self with the mobile rhythms of a radically negative space. Irrespective of disciplinary intentionality—indeed, apart from a somewhat vague nod towards something loosely referred to here as "deep mapping", how exactly might, or should a project such as this be labelled?—what it does point to is a praxis of the self as a fundamentally creative act: an embodied practice of what the geographer David Crouch [35] refers to as "spacing".

4. Rhythm Mapping: An Exercise in Composition

Perhaps inevitably, the Ballardian backdrop to a discussion that fixes its attention on a concrete traffic island leads us at some point to consider the phenomenological significance of *boredom.*

---

[2] The film draws much of its inspiration from the real-life case of the Frenchman, Jean-Claude Romand, who, like Vincent, was a Mr. Ripley character living a life of deception, pretending to his family and friends that he worked as a doctor and researcher at the World Health Organisation, while in reality spending his days driving aimlessly around the Alps. Romand somehow managed to sustain this for 18 years, his otherwise normal middle-class lifestyle financed by money embezzled from friends and relatives. When, in 1993, Romand's double life was finally exposed, he responded by murdering his wife, children and parents, and by attempting to take his own life [32].

Boredom does indeed come with the territory[3]. To briefly return for a moment to the M25, in Iain Sinclair and Chris Petit's 2002 experimental documentary *London Orbital*, boredom becomes a threshold that is there to be crossed; the M25, a "doorway into another reality". For Sinclair and Petit, the more that one is drawn in to the motorway's almost hallucinatory force field, the more it appears to exude its own consciousness. The goal of the motorway *flâneur*—whether artist, photographer, filmmaker, cartographer, anthropologist, psychogeographer, "motorway shaman" (or combination thereof)—is to in some way access this affective space (and in so doing help cultivate it). Deep mapping or deep topography comes to resemble a form of meditation practice:

> [T]he M25 is negative space—an energy drain. To enter it is to enter dead time; clockwise or anti-clockwise. More than other motorways, the M25 is designed to test the thresholds of boredom. It eliminates any romantic notion of boredom, but for the addicted it has its attractions: it is mainline boredom, it is true boredom, the quest for transcendental boredom...
>
> (narration by Chris Petit in *London Orbital*)

Whether transcendence can ever meaningfully be achieved is another matter. What motorway boredom inevitably leads to is perhaps an integral part of the journey itself, even if this is just the eventual exit up a slip road and a welcome retreat to a less transcendent geography of roundabouts, crossroads, and traffic lights ([15], p. 57; [38]). Towards the end of *London Orbital* we learn that one of the filmmakers' fellow travellers, the artist John Sargeant, fails in his own endeavour to transcend the limits of motorway boredom. Having retreated to a concrete island interchange to observe the road, his quest ends in inertia and disillusion: "After running too long on empty, everything stops—boredom, like rust, becomes entropy", remarks Petit, by way of rumination.

London Orbital is a film whose aesthetic and formal properties are well in keeping with its subject matter. The surveillant gaze of the moving image—the "flat literalness" of a digital journey through negative space—feels as hypnotic as the experience of driving (or being driven) through the abstract spaces of the motorway itself. The film sears its unrelenting and driving rhythm on the mental landscapes of the "voyager-voyeur" ([39], p. 105) in such a way as to slip, momentarily, into the experiential flux of actual motorway driving (and vice versa—the cinematic experience conjuring an embodied memory of motorway driving spaces). It is for this reason that *London Orbital* has never felt far from my peripheral vision as I have set about conducting my own enquiries into negative space. The film is also very much the product of two interlocking but

---

[3] Edensor cites the results of a report conducted into rail commuting, noting that only two percent of respondents found travelling boring. Edensor comments that "[T]he automaticity commonly ascribed as boring, allows commuters to dream, read, telephone and plan..." ([36], p. 198). This dialectical understanding of boredom points to the ways that boredom sets the experiential conditions for its alleviation through the activities it engenders. This is a similar approach to that which underpins the idea of non-places harbouring the seed or potentiality by which placelessness is negated and "place" (re)established through practice. Similarly, from a Lefebvrean perspective, the "felt experience of dullness, boredom and estrangement", as Rob Shields notes, is seen as an "[important] source of Utopian inspiration and revolutionary resolve" (in Fraser [37]). For discussion of the imbrication of the literary and the everyday in Lefebvre's work see Ben Fraser's *Toward an Urban Cultural Studies* ([37], pp. 95–120).

otherwise separate spatial methodologies: Petit's ruminant journeys around the motorway by car; and Sinclair's more peripatetic adventures on foot. If the abstract rhythms of the former have felt more closely in sync with my own experience of motorway driving (the journey up and down the M53 is part of my regular commute to and from work), the latter—a form of motorway navigation that deviates quite sharply from what might be regarded as routine—taps into rhythms that are closer to those encountered in my time spent on the motorway island.

There is, perhaps, an inclination to think of motorways as in some way neat, ordered, and "smoothly" configured to efficiently accommodate the unimpeded flow and circulation of traffic (even if this functional imperative routinely falls short of motorists' expectations). While motorways of course do exhibit these qualities it only takes the briefest of "off-road" diversions to access an altogether different landscape, one that is less ordered, more feral, and seemingly untroubled by the meshworks and desire lines of human activity. Similarly, when thinking about the kind of rhythms we might associate with motorways (from the mobile perspective of driver or passenger) it is those that map on to the experiential geography of the journey itself that are of significance: the *transito*[4] rush of forward momentum; the frustration induced by traffic jams or of slow crawls along congested carriageways; the violent collision of metal on metal; the parallel tracking of musical rhythm from the car radio or CD player ([36], p. 198; [41]). What is less likely to draw comment is the slow, arduous, and often painful progress of the body as it hacks its way through a dense thicket of trees and undergrowth, a growing blanket of dusk prompting a briskness of pace.

Nature, left to itself, makes no concession to the human visitor. This is an entry from my field notes, written shortly after my arrival on the island. *Very dense vegetation—difficult to move through island to the north, flatter area*. Boredom had little purchase in this environment, at least in these early stages of my residency; there was simply no time to be bored. The practical demands of orientation and of finding a suitable location that might serve as base camp (a clearing flat and comfortable enough to accommodate a body in a sleeping bag) imbued the landscape not with inertia but character and form. This was no negative space; no anomalous feature on the map. In the first instance it is an island woodland. Screen out the traffic noise and the fact that it was surrounded at all sides by motorway and there is little to distinguish it from other wooded areas that are located "inland" off the motorway.

The "north, flatter area" towards which I was headed is, in fact, not north at all, but west of my initial location. The directional geometry of the motorway (northbound and southbound) had imprinted itself on my cognitive map of the island in such a way as to streamline space and remove all but the most abstract of geographical forms. The non-place of the map had territorialized a space that in all other respects was anything but uniform or geometrically defined. This spatial confusion would become all the more apparent after I had spent a night on the island. The "north,

---

4 *Transito* is a concept that Guiliana Bruno adapts from the Italian philosopher Mario Perniola. This theorizes the spatialization of desire as inscribed in physical and mental motion, incorporating states of transit, transition, movement and circulation. With its emphasis on flows, movements and the spatial enactment of desire *transito* addresses itself to *transgressions* of social, spatial and moral boundaries, and the desires and kinaesthetic pleasures of embodied perception (see [40], p. 222).

flatter area" was where I would set up camp and initiate my nocturnal watch over the now moonlit and street lamp-bathed island (Figure 1). Come morning, when my exploratory adventures would begin proper, the sense of my being directionally-challenged soon became clear:

> Major orientation problems...I thought I was heading northwards [west] on the exploration this morning, but was in fact heading south [east]. Confusion over which side of the motorway I am on. I thought I was on the opposite side from the point where I landed but I am in fact on the same side, facing towards Hooton and south, not northwards.
>
> (Field note extract)

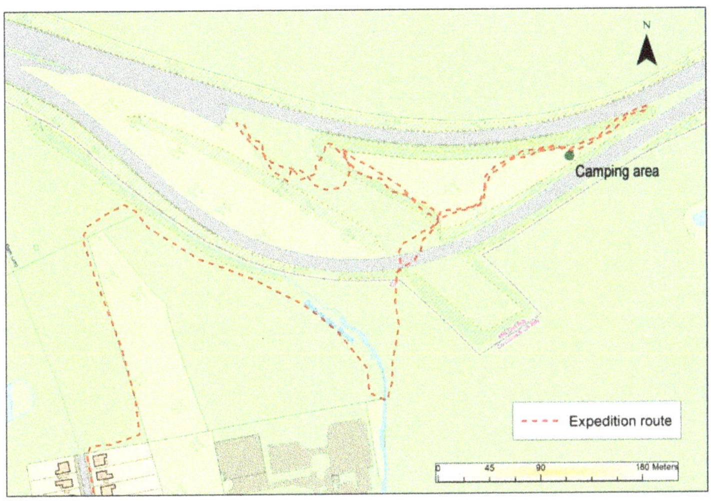

Figure 1. Ordnance Survey (OS) map of island showing expedition route, camping area and crossing points (1:2196). (Map data: © Crown Copyright and Database Right [2013]. Ordnance Survey (Digimap Licence)).[5]

This field note entry was recorded more than two hours after I had reached what I thought had been the most northerly (westerly) tip of the island. Lack of sleep, coupled with a befuddled sense of space, had served to unstitch the territory from its geo-locational points of anchorage, transforming it into a residual product of my ever so slightly wired imagination.

> Hubcaps, coke cans, an exhaust cylinder, half-brick, empty bag of garden chippings, large metal frame—possibly part of a sign, Quaver [crisps] packet, length of polystyrene coving, take-away coffee cup with lid, traffic cones (x 2), seat of office chair, rusty metal rope (with noose).
>
> (Field note extract)

[5] All images and audio–visual data are copyright of the author unless otherwise stated. In addition to the embedded visual content, hyperlinked digital materials can be accessed in the form of audio sound files (see Figures 5 and 7 and [42–43]) and a video shot during the period of fieldwork on the island [44].

The carpet detritus that littered the floor of the island—the sole vestiges of a proximate human presence (Figure 2)—contributed to the general sense of dislocation I was experiencing. The coming to rest of objects that had been washed up at random moments helped define an island ecology that has developed in isolation from the more manicured and "legible" landscapes of the motorway and surrounding area. In other words, despite the relentless soundscape of a chaotic and busy world "out there", the topographic texture of the island conspires towards an interiority of place and experiential dwelling. In an environment such as this the mind is left to follow its own course: consciousness and geography alike become unmoored. The sedentariness or restricted spatiality of the body provide fertile ground for the imagination to cut loose and take flight.

Figure 2. Hybrid textures.

I cannot recall now exactly at which point the significance of the soundscape began to leave its mark, but the challenges encountered in terms of spatial awareness were more than compensated by the cartographies of sound that became the default sensorial mechanism by which a habitual and "grounded" sense of space and place was provisionally secured. What might be understood as a "map" of the island non-place was thus one that derived its form from the pulses and sonic rhythms established by the unremitting flow of vehicular traffic. In retrospect, these island rhythms probably began to fully seep their way into the sensorium of my body in the early hours of the morning when I had bedded down to what would prove to be an extremely fitful sleep. The following field notes, written at 10:40 p.m. as I attempted to make myself comfortable for the night, provide some indication of the experiential space that had begun to take shape around me:

> Lying on mattress, blow up pillow. Looking up at the canopy and stars beyond. Full moon to the east [north]. As cars go by light ripples through leaves of the canopy. Sound of cars, like waves—irregular rhythm. Unwavering, few variations (bar the odd motorbike or lorry). They seem to have a life/biorhythm of their own, as if they are part of a bigger entity. Feeling of isolation—each one a world unto itself, surrounding my own isolated space...I wonder if the relentless hiss/rush of cars will have an effect on my sleeping/dreaming state (if I manage to sleep...).
>
> (Field note extract)

At 4:17 a.m., I am awoken by the sound of a tin can, thrown from a car, rolling along the motorway. No doubt, in time, this too would become a constituent part of the island's alluvial terrain: hybrid flotsam that, like me, finds its place in the entropy of motorway edgelands.

In a space—or non-place—such as this, the night waves and rhythms can cast something of a spell (Figure 3). The traffic noise—the soundscape—seems louder and more prominent than during the day (lorries and heavy goods vehicles make a particularly thunderous impression). As a consequence, it becomes all the more enveloping; its peaks and lulls secrete an inexorable pattern that lodges itself in the penumbral zone between consciousness and slumber. Despite its heightened volume, the traffic may even have had something of a soporific effect, although this is less easy to recall with any degree of accuracy. Indeed, as in autoethnographic terms I had become, for three or four hours at least, an "unreliable narrator", it is the sound recordings I had been making since my arrival on the island that provide the sole means of documentation during the few hours of sleep.

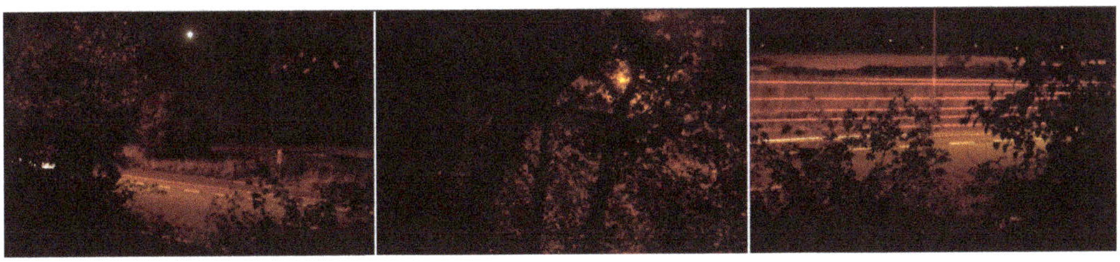

Figure 3. Night waves.

Insofar as the all-embracing soundscape so clearly defines the embodied spatialities of the island, it is the frequency, intensity and rhythmicity of the traffic noise that provide the raw material from which to best engineer a cartographic "capture" of this experiential space. The time domain images extracted from the sound recordings (see Figures 4–8), if processed in conjunction with the audio recordings themselves (see links to audio clips—Figures 5 and 7), render possible a form of locational positioning whereby, representationally at least, the island may be re-visited or re-inhabited and a rudimentary degree of orientation established. As maps, the sound capture images delineate the temporal geographies of the island in ways that give an impression of the rhythmic modulations that both anchor the space in the durational flux of phenomenological time and track the nocturnal and diurnal rhythms of a social ecology built on mobility and flow. This latter, more cyclical space-time transcends the time-stamped singularity of the captured moment (the contingent occurrence of my "being there") to form the entrenched social rhythms that are a constellated product of the myriad "spatial stories" [7] and everyday mobilities of the motorway wayfarer. Accordingly, as Lefebvre suggests, "[t]he cyclical is social organisation manifesting itself. The linear is the daily grind, the routine, therefore the perpetual, made up of chance and encounters" ([2], p. 30).

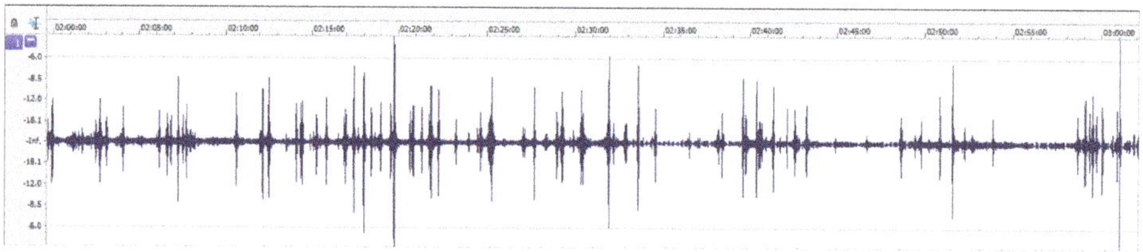

Figure 4. Soundscape map, 2:00 a.m.–3:00a.m., 23 August 2013 (53°18'10.1"N 2°58'00.9"W).

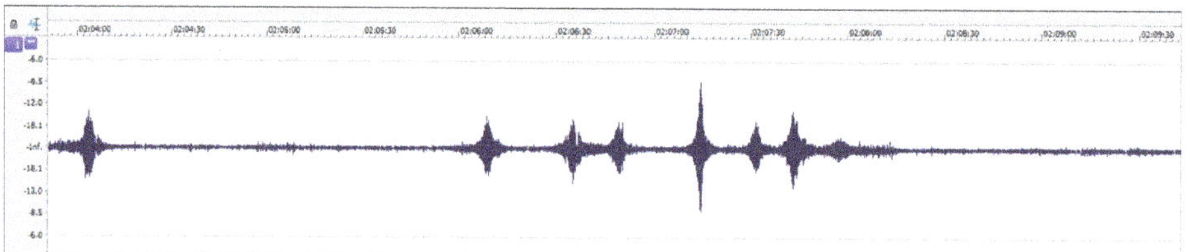

Figure 5. Soundscape map (detail), 2:04 a.m.–2:10a.m., 23 August 2013 (53°18'10.1"N 2°58'00.9"W).To listen to soundscape see: [42] (recording best heard with headphones).

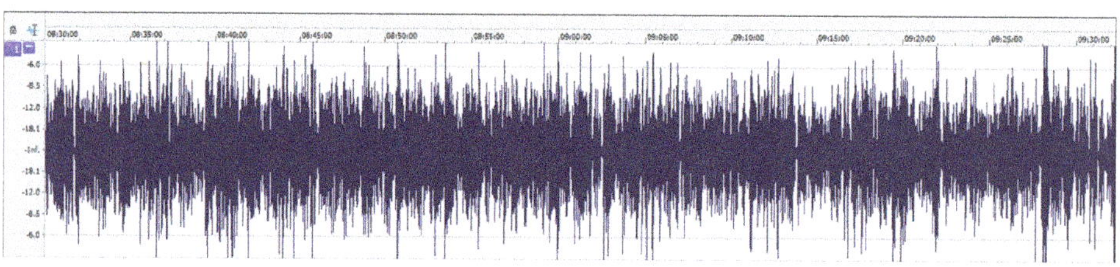

Figure 6. Soundscape map, 8:30 a.m.–9:30a.m., 23 August 2013 (53°18'10.1"N 2°58'00.9"W).

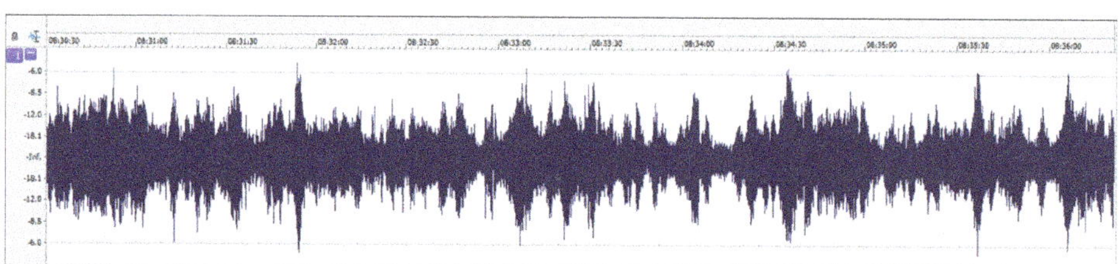

Figure 7. Soundscape map (detail), 8:30 a.m.–8:36a.m., 23 August 2013 (53°18'10.1"N 2°58'00.9"W).To listen to soundscape see: [43] (recording best heard with headphones).

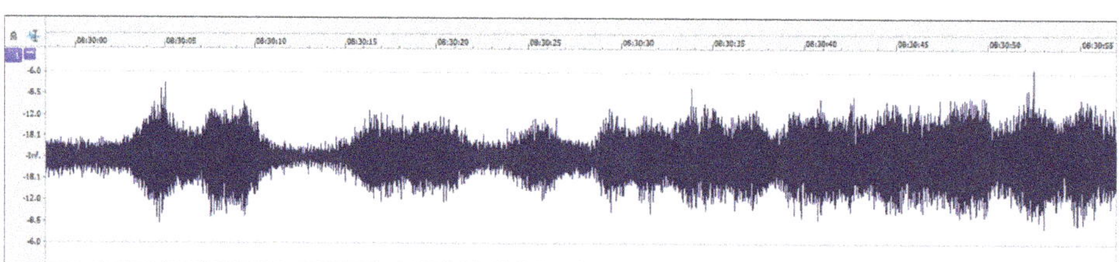

Figure 8. Soundscape map (detail), 8:30 a.m.–8:31a.m., 23 August 2013 (53°18'10.1"N 2°58'00.9"W).

If we compare Figures 4 and 6 (2:00 a.m. to 3:00 a.m. and 8:30 a.m. to 9:30 a.m., respectively), the difference in the intensity of rhythm is stark and vivid, not all that surprising given that one of the maps corresponds with the height of morning rush hour. From one vantage point, then, the rhythms of the road represent a chronometric reading of a purely functional, processual and utilitarian space. From another, however, and one more in keeping with the idea of the island as an embodied and experiential space, these representations map onto an interiorized world of which, as "rhythmanalyst", I am both monarch and intrepid explorer. Revisiting the space–time of the island through sound, the images can quite readily be re-imagined as electroencephalographic readings of brain wave activity. Between 2:00 a.m. and 3:00 a.m. it is a landscape shaped as much by gaps and lacunae as the indicative verticality of "events". The mind is wound down and given over to the contemplative or somnolent flux of phenomena that, come morning, has succumbed to the more fevered density of ambient activity. By contrast, between 8:30 a.m. and 9:30 a.m. there is little in the way of horizontal "space" in which the mind can wander. Instead we are thrown headlong into the fray of a dense conurbation: a surging cluster of individuated time that barrels across the page like a juggernaut.

In his "previsionary portrait" of the rhythmanalyst, Lefebvre presents us with a figure who "listens—and first to his [sic] body; he learns rhythm from it, in order consequently to appreciate external rhythms. His body serves him as a metronome" ([2], p. 19). The metronomic function of the researcher's situated body is a central methodological component that marks out her or his practice from that of, say, the surveyor or topographer. The body *feels* for rhythm, a process that allows for the spiralling out of embodied time to the spatiotemporal precincts of the wider social world, a conjunction that can be felt in either resonant or dissonant terms (or, as is more likely, both). Lefebvre goes on to remark that the body produces a "garland" of rhythms: "the surroundings of bodies, be they in nature or a social setting, are also *bundles*, *bouquets*, *garlands* of rhythms, to which it is necessary to listen in order to grasp the natural or produced ensembles" ([2], p. 20, emphasis in original). The time domain images, especially those extracted from sound recordings made during the day, might themselves be likened to garlands inasmuch as they represent a bundle of temporal moments "strung together" in the fashion of what Lefebvre describes as "an aesthetic arrangement" ([2], p. 20). Garlanded by rhythm on all sides, my time on the island was not entirely my own insofar as my embodied self was in part a composite assemblage of the time of others. But equally, by adorning myself in the garlanded rhythms of the motorway I was able to cultivate and fashion my own spatiotemporal corporeality. Bringing into sharper relief the very *otherness* of the rhythms that defined the space-time of my island dwelling, the marooning helped make possible a *time-unto-myself*: an experiential "time out" that provided the space for a creative re-ordering or re-assemblage of the space–time rhythms of an otherwise mundane roadscape. Viewed methodologically, the function of the body in this non-place setting is, therefore, that of a rhythmanalytical apparatus or interface: the body as a *site* of spatiotemporal praxis.

5. Road to Nowhere: Beginnings

If the discussion has thus far centred more squarely on temporality and rhythm in terms of an embodied mapping of the motorway island, no less important are considerations of its material geographies and why it is that such a topographic anomaly is there at all. However seductive it may be to think of this space as existing "outside" of the quotidian trammels of history (as if a "bubble" of *utopic* non space that has somehow fallen between the cracks of Euclidean social reality), a more prosaic reflection would be to cast attention on the historical geography of this stretch of the M53 motorway and the circumstances surrounding the island's formation. Refer to any map showing the area just to the west of Junction 5 on the M53 and it is not difficult to infer an obvious design intent on the part of the motorway engineers despite the fact that these plans never came into fruition. A provisional history can be gleaned from even the most cursory cartographic scrutiny. On a satellite image, for example, a curious oblong feature seemingly protruding from the island on the south side of the northbound carriageway has obvious affinities with the island topography in ways that cannot fail to pique geographical interest (Figure 9). If we turn to a detailed Ordnance Survey (OS) map of the island the relationship between the two features is rendered all the more apparent (Figure 1). That this island protuberance is the result of an aborted motorway design can be confirmed by reference to a 1970s OS map published when the M53 was still under construction. The oblong is now clearly understood as being part of a planned extension of the motorway that was to have continued in a south-westerly direction (Figure 10). The clincher is determined by reference to one final map showing the newly constructed motorway and road network around the Junction 5 exit (Figure 11). The triangular shape that defines the area that has since come to form the island is made up of what were to have been two slip roads exiting the motorway that would eventually link up with the M531 that bypassed Ellesmere Port to the east. The carriageway of the M53 was to have continued south thus forming what is now the bulk of the island terrain as well as the adjacent patch of land. What were to be the slip roads are now the south- and northbound carriageways of the M53, which continues southeast along the route of the former M531. In other words, to position oneself on the island is quite literally to be on a road to nowhere. A more visible monument to this short-lived plan was a flyover bridge that, until its demolition in the early 2000s, spanned little more than the width of the motorway's northbound carriageway (Figure 12).

Work on construction of the M53 Mid-Wirral Motorway began in 1969 and was carried out in various stages until 1982 when the last section was opened, extending the motorway to the outskirts of the city of Chester. The M531, which had been extended to join up with the M56 North Cheshire East-West Motorway, was completed in 1981 and would became part of the M53 as it exists today, which runs along the spine of the Wirral from the Kingsway (Mersey) Tunnel in the north to the A55 North Wales Expressway near Chester in the south ([46], pp. 311–13). In their report published in 1966, the Consulting Engineers commissioned by the Ministry of Transport to carry out a location study for the new motorway recommended that the route be extended south to the interchange at Dunkirk with the A5117. This would have linked with a newly extended M56, but also would have provided a faster route for traffic bound for North Wales along the A494 ([47], p. 10).

Figure 9. Aerial view of island (Map data: © Google, Bluesky, Getmapping plc [2015]).

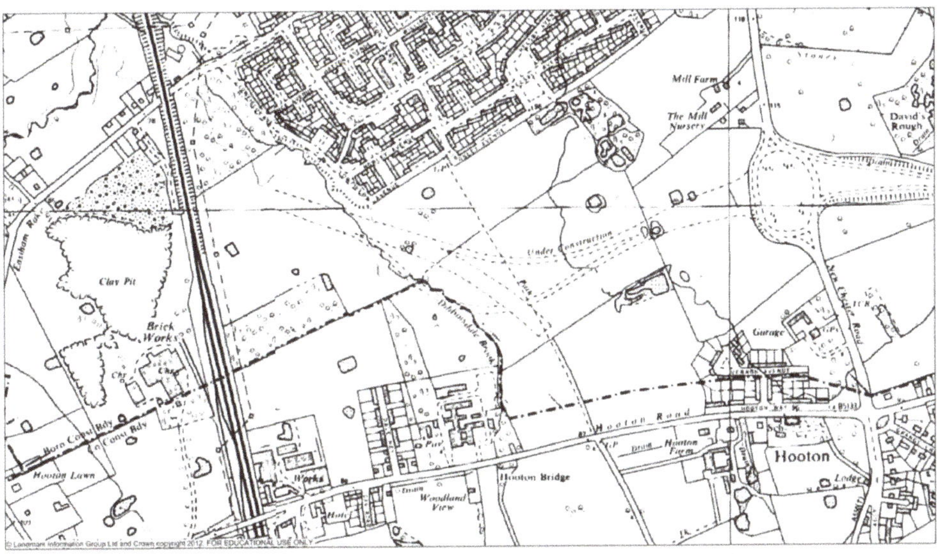

Figure 10. 1970s OS map of island (under construction) and surrounding area (1:7500) (Map data: © Crown Copyright and Database Right [2013], Ordnance Survey (Digimap Licence)).

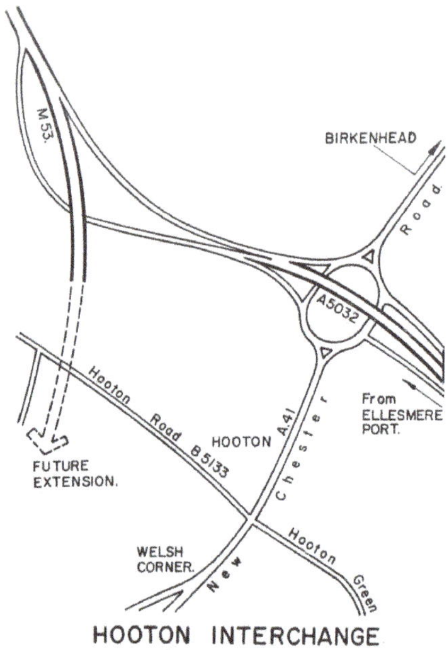

Figure 11. Illustrated map showing proposed southbound M53 extension. (Department of the Environment [45], © Crown Copyright (n.d.)).

Figure 12. Aerial view showing motorway bridge built as part of proposed southbound M53 extension, c. 2000 (Map data: © wikiwirral.co.uk [2008]).

As already mentioned, the M53 is a road that I use often as part of my regular journey from North Wales, where I live, to my place of work in Liverpool. Junction 5, which is near to the location of the island, is the point at which I enter and exit the motorway. Had the proposed southbound extension gone ahead this would have been the route I would have routinely taken (cutting a good ten minutes off my commute time in the bargain!). Given this, perhaps my "islandness" [48] might be construed as nothing more than a performative work-through of a latent geography that was destined not to be consummated (a utopian imagining sprung from an unrealized sense of auto-mobile satisfaction). However, a more likely rationale would be one that sought to join the dots between an experience of motorway "driving space" [20] and the "off-road" geographies that, over time and repeated iterations, have become firmly imprinted on my cognitive and affective map of the M53 as an everyday landscape. Because the distance travelled along the motorway is relatively short (the M53 is only 19 miles from beginning to end) it has rarely felt as monotonous a driving experience as that encountered on, say, the M6, or other parts of Britain's motorway network. Edensor's observation that "confrontations with forms of otherness [such as fantasies, disruptions, lines of flight, sensual intrusions]...disrupt routinized experience and practice, throwing into sharp relief constructions of normative automobility" ([29], p. 155) does not adequately take into account the extent to which such disruptions are themselves part of routinized and habitual experience. The question: "How might routine commutes by car on the motorway be defamiliarized?" ([29], p. 155) can thus partly be answered by pointing to the ways that commuters *routinely defamiliarize* their surroundings through recourse to the flights of imagination, disruptions, sensualities, and so on that are part and parcel of the negotiation of these and other everyday landscapes. To reiterate the argument made earlier: non-places contain within them the potentiality of their own negation; motorways, on account of the very mundanity and featurelessness that Edensor's argument sets out to counter, are affective spaces given over to the "commingling" ([35], p. 12) of the self with the entropic energies of a landscape in flux.

One of the (de)familiarized features of the motorway which I found myself becoming ever more entangled with was the verdant mantle of trees that defined the territory that I only later realized to be an island. Approaching along the southbound carriageway, the apparently pointless margin of hatched road markings that flank the northern stretch of the island, along with a red and white "road closed" sign erected to (presumably) prevent cars from stopping or using the road markings area as an additional lane, were what first drew my attention (Figure 13). As subsequent journeys and more attentive rubbernecking would make clear there is more to this landscape than meets the eye. For a start, much of the island terrain remains unlandscaped; it is a pocket of land otherwise untroubled by the upheaval of road construction, as if it had somehow managed to hold off the territorialising march of the bulldozer and excavator and the contractors had resignedly accepted that they would have to work around it instead. As we have already seen, aerial views and maps allow us to clearly make out exactly where the excavatory incision for the road-that-was-not-to-be had been made. But either side of this cleft remains a dense thicket of trees and topographically uneven terrain that thumbs its nose at the uniformity of the surrounding road. This was a landscape that had the appearance of concealing a story that it had never got around to telling itself. Beyond the piecing together of a truncated history from official archival records, or the online chatter of discussion forums where amateur historians or transport enthusiasts trade local knowledge and geographic speculation[6], the narrow curve of unkempt (and untouched) greenery offers itself up as the only viable resource from which to extract a semblance of a geographical or archaeological narrative. A hastily thrown glance as I sped by at 60 or 70 miles per hour could only disclose so much.

Figure 13. View approaching the island on the southbound carriageway (Map data: © Google [2014]).

6. Making Tracks: Embracing Islandness

> I hacked through dense thicket for a while then saw a ditch leading up/away [providing a route] with less dense obstructions. I hacked a clear path along this, making sure that the path could be easily identified for the return journey. I continued along this as far as I could before hitting an impenetrable wall of bramble and trees. I scrambled up a bank to the left of the "path" and, after some minor hacking, found this led out onto an open "meadow"...
>
> (Field note extract)

After my night spent under the orange-glow canopy of trees to the east of the island (near to the point where I had landed), and the spatial disorientation that marked, come morning, my initial attempts at exploration, the objective for the second day was to make my way to the western edge (although, in keeping with an internalized directional logic of northbound/southbound, I was still thinking of this in terms of "north"). The journey was one that mostly involved trekking through the interior of the island, away from the perimeter edges and the rush and immediacy of the morning traffic (a concern with being spotted by passing motorists, particularly lorry drivers who

---

[6] Research into the construction and planning of the M53 motorway was undertaken at Cheshire Record Office in Chester: [49]). Useful information was also obtained from a number of online resources, such as discussion forums focused on the history and heritage of Britain's motorway network [50–52].

had a higher vantage point, meant that the journey "north" offered a welcome opportunity to once again don a cloak of invisibility) (Figure 14).

Figure 14. Flow.

The difficulties encountered in navigating those parts of the island that have remained largely untouched by human intervention were compensated by the reward of arriving at the open expanse of grassland which (not inaccurately) I referred to as the "meadow". The meadow represents that area of land that had been carved out by the motorway contractors as part of the aborted plans to extend the route directly south (Figure 15). As my field notes describe,

> This open area is comprised of tall grasses, thistles, wild flowers, and a number of crumbling brown-earth ant mounds. Beyond the clearing is another dense mass of trees and brambles, which dips down into a "valley" type area. Beyond that the island flattens and thins out and comes to an end. I walked around to the edge of the last group of trees and could just make out the stripy road markings that mark the start of the island, but it was not possible to see these clearly without being seen by motorists. The clearing area, as it is open, is mostly in full view of the southbound traffic. The meadow/clearing must be the highest point of the island.
>
> (Field note extract)

By the time I reached the meadow it was midday, the sun had fully broken through the intermittent cloud cover, and a light breeze was toying gently with the tall grasses and ragwort. After some time spent photographing and videoing the landscape and flora, I sat for an hour or so soaking up the warmth and the undulant rhythms and succumbing to a feeling of general wellbeing and corporeal detachment. If I had been climbing a mountain this would have been the summit moment: a snatch of well-earned "time out" before the final descent. This part of the island was also the widest, meaning that I was furthest from the motorway than at any point in my residency. Because of this it was possible to get more of a sense of the landscape as existing on its own terms and not as merely an adjunct to the roadscape or as a negative space defined wholly in relation to the motorway.

An entry in my field notes written at 04:17 a.m., when ensconced in the wooded area that I had made my base for the night, recorded the following observation: *Appears to be little in the way of insects or wildlife*. Nor, from what I recall, was there much diversity in terms of plant life on this part of the island. It was as if the location's close association with the motorway, just a few feet to the north, had taken its toll on the island ecology and drained it of vim and vitality (something that, up to a point, resonated with my own embodied experience during this period). In the meadow, by contrast, life had all the appearance of being a good deal more abundant. While it would be premature to start imagining some form of utopian settlement eking out an existence on the island's untapped natural and aesthetic bounty, the affective appeal of the meadow does invite reflections on ideas of "islandness" conceived of as a spatial practice oriented in response to the mainline rhythms and spatiotemporal structures of everyday modernity.

Figure 15. Meadow.

In their self-styled "non representational" approach to the question of island dwelling[7], Vannini and Taggart argue that "islandness" is best approached not in abstract terms (*i.e.*, "what is an island?") but as an affective experience and practice ("how do you *do* your island?"). Viewed thus, islandness "changes from a representation inside our heads to a set of tasks unfolding in front of its inhabitants" ([48], p. 235). In other words, the ways in which an island is *dwelt upon* determines the meanings and shape that make the island what it is: "the life of your island is the sum total of the sensations it gives rise to, the cumulative incorporation of those feelings carved into its soils and shores, and the embodiment of its affective spaces on its dwellers" ([48], p. 236). Although such a formula might conceivably be applied to *any* type of landscape (it also raises the question of whether a performative/non representational understanding of islandness is necessarily predicated on there being something that topographically/representationally *is* an island in order for it to be valid) what it does usefully point to is the way islandness is carved out through the different forms of practice, movement and mobility that temporalize the landscape and confer on it the more anthropologically-loaded status of "taskscape".

For Ingold, the concept of *taskscape* offers a strategically important tool with which to re-think *landscape* in terms of its fundamental temporality. "[T]he temporality of the taskscape," he suggests,

> while it is intrinsic rather than externally imposed (metronomic), lies not in any particular rhythm, but in the network of interrelationships between the multiple rhythms of which the taskscape is itself constituted....the forms of the taskscape, like those of music, come into being through movement. Music exists only when it is being performed; it does not pre-exist, as is sometimes thought, in the score, any more than a

[7] For a critical discussion of perspectives on so-called "non-representational" geographies of place and mobility, and comparisons with approaches developed in phenomenological and existential anthropology, see Roberts and Andrews [53].

> cake pre-exists in the recipe for making it. Similarly, the taskscape exists only so long as people are actually engaged in the activities of dwelling. ([54], p. 197).

If we extend this reasoning to maps (as Ingold himself goes on to do in a subsequent chapter of *The Perception of the Environment*), and to the process of mapping the embodied landscape/ taskscape of the motorway island, then what we might understand as "deep mapping" acquires greater resonance insofar as it helps tease out the temporal dynamics ascribed to a cartographic method that is critically underpinned by an anthropological sensitivity to landscape as a fundamentally social, embodied and lived set of practices. Deep mapping is thus strongly consonant with an idea of "wayfinding": knowledge that is "cultivated by moving along paths that lead around, towards or away from places" ([54], p. 229). As a product of wayfinding, spatial and environmental knowledge is "ambulatory...[in that] we know *as* we go, not *before* we go" ([54], p. 230, emphasis in original). Mapping-as-wayfinding unborders the map as a frame or plane of representation. The cartographer-as-wayfinder, by extension, is a fundamentally embodied subject whose field of practice is similarly unbounded and more productively thought of as a *meshwork* in which she or he is reflexively entangled. Mapping, in this sense, is as much about the embodied self as it is about the space which the cartographer (or spatial anthropologist) is oriented towards as a focus of study and in which s/he dwells [55].

Dwelling in the midday sun of my island meadow, the less fevered rhythms of the road now more of a background tapestry of sound than in-your-face maelstrom, the landscape I had at my disposal was one that to a large extent had been willed into being. Even now, as I (re)construct the performative map of my island imaginary, the iteration of its fictive geographies is such that whatever might be gleaned in terms of "islandness" is nothing but the residual and contingent product of my being-there (an islandness, in other words, that can only ever be "my" islandness). Much of what would have passed through my consciousness, like the evanescent flow of traffic around me, has long since faded into oblivion. The traces that remain—whether hastily scribbled field notes, digital captures of various description, or memories that have coalesced into renderable form—have provided the basis for a tenuous archaeology from which I have crafted the outline of a narrative journey, the performance of which has given shape and substance to a "place" that is otherwise not a place (or that *is* a non-place, to affirm the negative). If I remove myself from the equation there is no island to speak of, merely an inconsequential pocket of land tucked between two motorway carriageways (two slip roads that were destined not to be) which, for much of the four decades or so it has existed, has probably mustered little if any human interest. It is negative space in the same way that the area beneath objects such as chairs or tables, made concrete (literally so) in work by the artist and sculpture Rachel Whiteread, is negative space. It is not that the space is not there but that it is through the intervention of the artist (or in my case the autoethnographer or spatial anthropologist) that what is not *not there* acquires a rudimentary form. To dwell in such a space is therefore to both conjure and embrace this form. Islandness is culled from abstraction in the same way that an imaginary is culled from the imagination. In *Dialectic of Duration*, Bachelard remarks that "The story of a journey is a function of its geography" ([4], p. 89). The geography of the island is that which, in the first instance, makes it an island (the liquid space of the motorway replacing that of water). But the island's geography is also what makes it an island in the narrative sense (a story and performance, an academic fiction). Islandness is the story of a journey. Moreover, it is in the journey that we trace and retrace what it is that islandness might conceivably delineate at any given juncture.

Mapping the islandness of the island is a process of, firstly, *making tracks*. These are laid down in the act of walking, the gleaning of digital ephemera (images, audio, geospatial tracks[8]), and the embodied immersion in a space that is ordinarily marked out in terms of its placelessness and abstraction. The follow-through is the re-tracing (re-*making*) of these tracks and moulding them into a bricolage or assemblage, the function of which is to convey a spatial story that maps onto an embodied sense of place: the *being-there-ness* of my island dwelling. The task of re-tracing/re-making is constituted in the process of writing-up. In this respect the relative "depth" of mapping is a measure of just how vivid or impressionistic the performative space of islandness actually is in terms of a cartographic imaginary. As an autoethnographic narrative, the locative properties of the text are such as to situate myself within a representational space that both frames my "being-there" (the recounting of a *journey* in the etymological sense of a "day's work or travel" (from *jornee* in Old French)) and creates a setting (a *mise-en-scène*) in which to explore the imbricated rhythms and temporalities by which "islandness" is set adrift and plugged into a wider matrix of space, time, and everyday mobility.

As a purely geo-locative spatial story—a meshwork of lines digitally inscribed on a map—islandness can be textually reconstructed in the form of an itinerary or journey: a traversing of physical space. But this too can be shown to register the pull effect of centrifugal space, the affective force of the motorway's hyper-mobile and "placeless" environment threatening to destabilize an otherwise Euclidean pattern of lineation. The maps shown in Figures 16–19 each chart—up to a point—the route I followed throughout my time on the island. I digitally recorded the tracks of my journey using two separate devices: a GPS-enabled tablet computer and a Garmin eTrex GPS receiver. Both sets of tracks are represented on the map at scales of 1:5500, 1:1500, 1:100, and 1:10. In Figure 16 the whole route is visible, from the starting point at a residential street called Dale Hey where I had parked my car, to the point where I crossed over the motorway (a journey which in itself was by no means straightforward, having been determined as a result of two prior recce, or reconnaissance, visits), and then my meanderings across the island itself. The return journey and crossing followed the same route.

However, where the GPS signals dropped out momentarily en route there are several spikes that are visible, the most prominent being the red (tablet) lance-like feature jutting northwards and its black (eTrex) counterpart pointing south-east. If we zoom in to the next scale (Figure 17) we start to get a clearer picture of just how erratic and undisciplined the track lines in fact are. Zoom in again (Figure 18) and a thick knot of red lines reveals a quite alarming degree of agitated movement around what is less than a 100 square meters area to the east of the island. This surely cannot be an accurate representation of my actual spatial mobilities. If it is, then I am at a loss to account for the frenetic rhythms being tramped out or why it was I thought that such a complex and baffling pattern of movement might have been necessary. If we zoom in one final time (Figure 19),

[8] Digital content (audio soundscapes and video) can be accessed via hyperlinks provided in note 5 above [42–44].

then it is almost as if we are entering a different spatio-temporal dimension entirely. If evidence were needed that, at a micro level, some other, wholly unaccountable form of spatial activity was being tracked then this is it. The area shown on the map is little more than five square meters. The geometric complexity revealed at this scale maps a web of mobility that resembles nothing that is corporeally human. In fact, the points on or near the island where these disturbances mainly seem to occur correspond with those where movement was minimal, such as the location at the edge of the motorway where I waited to cross over to the island. The dense mass of tangled lines to the east of the island marks the location where I had camped for the night. If these patterns represented the tracking of movements in space then they were not any that I could convincingly lay claim to.

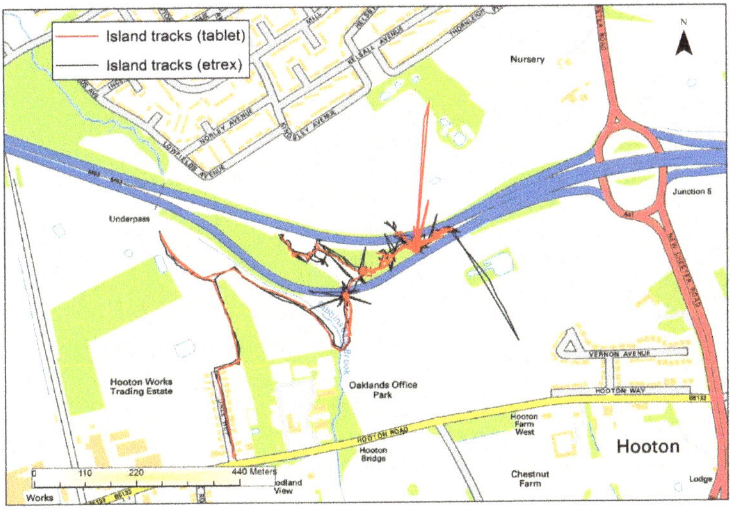

Figure 16. Map showing GPS tracks of island field visit, 22–23 August 2013 (1:5500) (Base map data: © Crown Copyright and Database Right [2013], Ordnance Survey (Digimap Licence)).

Whatever technical quirks or environmental anomalies these tics and spikes might be attributed to lies beyond the scope of the present discussion. However, given that they tended to occur at locations where I was mostly stationary it is hard not to equate them with the affective traffic of thoughts, sensations, rhythms, reveries, flows, transmissions, and general mental meanderings that arise when the body comes to rest. Trading the "taskscape" of the purposive trek (track lines with few aberrant deviations) for the unfettered space of the imagination (track lines that display no discernibly rational spatial logic), the body's entanglement with this meshwork of lines opens it up to a qualitatively different set of mobilities. As with the soundscape maps discussed earlier, rhythms captured as a result of empirical methods of environmental data capture function instead as *detourned* mappings that project a purely experiential and embodied sense of space. The pregnant liminality of this space allows for the steady accretion of temporal topologies ("time out") that deterritorialize and "garland" the embodied self. The centrifugal pull of the motorway exerts a force that suffuses almost every facet of island dwelling, even—or rather especially—in those moments when consciousness takes flight far beyond the space–time of the present.

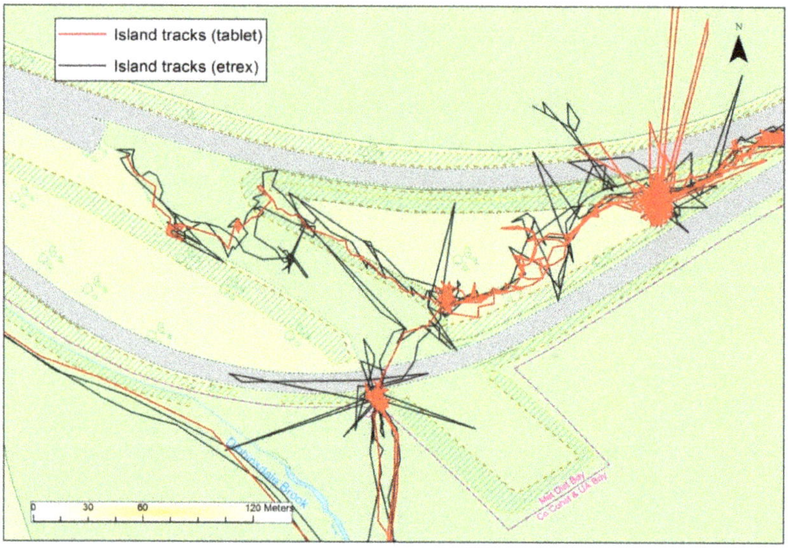

Figure 17. Map showing GPS tracks of island field visit, 22–23 August 2013 (1:1500) (Base map data: © Crown Copyright and Database Right [2013],Ordnance Survey (Digimap Licence)).

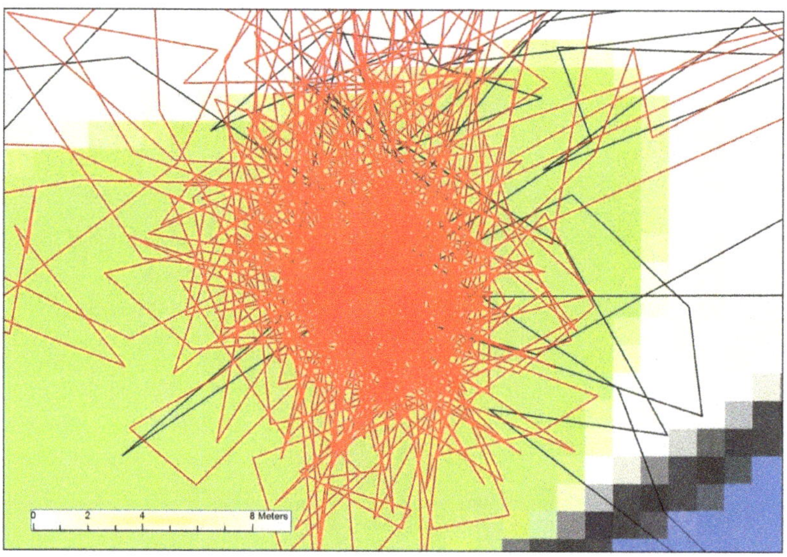

Figure 18. Map showing GPS tracks of island field visit, 22–23 August 2013 (1:100) (Base map data: © Crown Copyright and Database Right [2013], Ordnance Survey (Digimap Licence)).

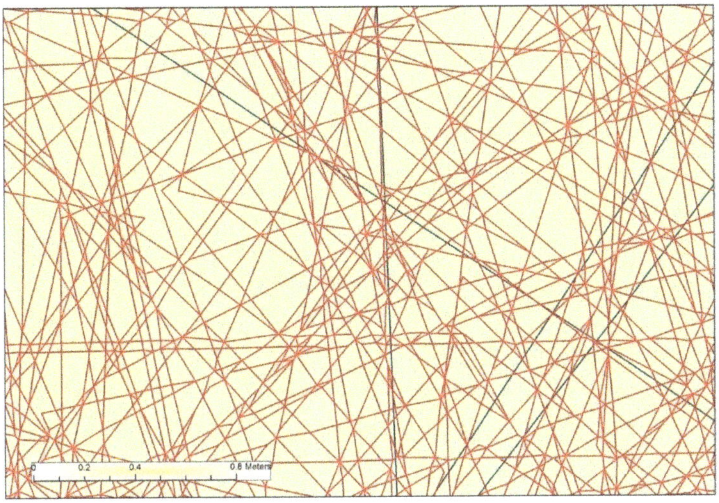

Figure 19. Map showing GPS tracks of island field visit, 22–23 August 2013 (1:10) (Base map data: © Crown Copyright and Database Right [2013], Ordnance Survey (Digimap Licence)).

7. The Rhythm of Non-Places: Exit.

When it came time to make my departure from the island there was already a palpable shift in rhythmicity, back towards the more routinized temporalities encountered as part of my everyday spatial practice. My crossing back over the carriageway, while executed with the same degree of caution with regard to personal safety, was characterized by a more lackadaisical attitude when it came to being seen. For some reason this now seemed less important, and as I scrambled back up the embankment that led to the farmland I had crossed over the day before I realized I was in full view of the flow of traffic that had resumed along the stretch of the motorway that skirts the southern flank of the island. Perhaps a feeling of closure and accomplishment had started to kick in. Perhaps my fatigue and aching limbs (not to mention lacerated forearms from all the hacking through undergrowth) had begun to numb the excitement and sense of danger, blunting the edge of the island's former rhythmicity. Perhaps I was just fed up and in desperate need of a shower. Whatever the circumstances, approached in terms of its spatio-temporal *rhythms*, what makes the island what it is (by which I mean what gives it its "islandness") are precisely those temporal geographies which, upon arrival, disrupt the habitus of self and body, and which, upon exit, are experienced less acutely.

Rhythms, as Edensor observes, "are essentially dynamic, part of the multiplicity of flows that emanate from, pass through and centre upon place, and contribute to its situated dynamics" ([3], p. 3). The rhythms that make the non-place of the island what it is are, of course, particular to its own situated dynamics. They are also particular to my own situatedness as an embodied subject, rhythmanalyst, autoethnographer, (deep) cartographer, or whatever other label we might choose to affix; the point being that the institutional and individual habitus of the body plays a key part in determining, feeling, intuiting, interpreting, mapping, collecting, assembling, or *listening in to* the rhythms that constitute the island's observable islandness. Gottschalk and Salvaggio argue that non-places exhibit "distinctive temporal parameters that replicate, deviate from, or completely subvert those that typically organize everyday life" ([56], p. 16). While this is borne out by the analysis presented in this study, the timbre and rhythm that makes these temporal parameters distinctive are by no means uniform. The distinctiveness of non-places is not just on account of their constituent temporalities, but also of their heterogeneity. Although, by definition, non-places lack the social and symbolic embeddedness of anthropological places (and hence why they are typically defined in terms of their *lack* of heterogeneity and identity), when interrogated with the depth and rhythmanalytical precision required to draw out the temporal affects of these spaces, or, when recognising that "the *soundscapes* of non-places are also important stimulators of movement, interactions, and subjectivities" ([56], p. 27, emphasis in original), the extent to which there *is* often a distinctiveness and heterogeneity to speak of becomes that much more apparent.

In this respect, a titular focus on the "rhythm of non-places" should not be construed as an attempt to refine a spatial taxonomy to which I am ascribing a transferable set of "defining" characteristics. Such an objective would be almost as reductionist and questionable as that which set about exploring the "rhythm of *places*". Which places? When? Whose rhythms? How are they measured? To what ends? These are all questions that can just as readily be posed of *non-place*s. Instead of subscribing to an excessively rigid application of the concept (at the expense of fine-grained, ethnographic and performative insights) a more instructive approach is to acknowledge the dialectical configuration that makes spaces such as a motorway island what they are *in relation to* places that are routinely encountered (and thus more likely to be invested with symbolic meaning and structures of identity). The question that is jostling for attention is not what makes places non-places (a rather well-trodden and near-exhausted line of enquiry) but rather what it is that makes non-places *places*. To which, of course, the provisional answer is: people.

In epistemological terms, the task of populating non-places is one that informs the necessary (if still nascent) development of a spatial anthropology; in other words, a closer degree of engagement with those who move through, inhabit and produce such spaces. By extension, methodologically the task becomes one of embracing ethnographic and autoethnographic approaches: populating in the sense of the performative emplacement and direct intervention of the researcher in the field. In my case, this latter extrapolation brings with it a more literal connotation in that I *was* the population where before there had been none. Autoethnography, and an analytical gaze turned partly inward, becomes the default setting: there simply was no one else on the island—no person Friday—to whom I could turn my attention or share my rhythms. The islandness or liminality I experienced was not, therefore, shaped by a sense of communitas [57] or intersubjectivity [58], but rather by the anonymity and abstraction of the hyper-mobile environment in which I was marooned. In such a setting, my *self* and my *body* constitute no less of what counts as the "field" as the taskscapes of the island itself:

> The rhythmanalyst calls on all his [sic] senses. He draws on his breathing, the circulation of his blood, the beatings of his heart and the delivery of his speech as *landmarks*...He thinks with his body, not in the abstract, but in lived temporality. ([2], p. 21, emphasis added).

An enquiry into the rhythms of non-places is, therefore, at the same time an enquiry into the rhythms of the self. Thinking with her body in lived temporality, the rhythmanalyst of non-places maps resonant and dissonant connections with the ambient space-time "noise" around and beyond her immediate field setting. Proceeding from the existential and "radically empiricist" [59–61] grounding of a field practice attuned to the micro-sensorial and entangled commingling of lived space and embodied time, the rhythmanalytically-observant anthropologist variously reads, feels, senses, captures, measures, channels, conducts, *detourns*, visualizes, represents, gives vent to, and ruminates on the rhythms—the lived temporality—in which she dwells. One of the key challenges posed by an anthropology of non-places, however, is finding ways to feed out to (and back from) a wider social and intersubjective field of mobility and practice. In other words, how to go beyond the specifically *auto*ethnographic remit of the research. Again, this hinges on the specificity of the non-place in question. An ethnography centred on, say, an airport [62] or the Las Vegas Strip [56], given the steady flow and throughput of people, would of course offer more opportunities in this regard. For a space such as a motorway island—not the kind of setting one would expect to run into prospective informants—the challenge is that much more difficult.

In this vein, the very last task on my island expedition consisted of fixing a laminated flyer onto a strategically chosen plank of wood. The wood fords a brook that marks a key entry and exit point to and from the island. The flyer features a design logo for the motorway project, which at the time I had been calling "Concrete Island". The flyer also includes the URL of a webpage that provides some basic information about the project along with contact details. This was in fact the last of several (identical) flyers I had erected, all but two of which had been pinned to trees at various locations around the island (Figure 20). As might be guessed, the point of the flyers was to keep open the possibility of contact with anyone who chanced upon these particular locations and who might be willing to be drawn into discussion about their own "being-there" (or, indeed, their own "islandness"). The two flyers I had erected at non-island locations (the border brook and a fence near to the point where I had crossed over the motorway) were unlikely to attract the attention of any casual passer-by. Given the difficulty in accessing these locations whoever stumbled upon the flyers would have to have had some purpose or design that had led them there. While there is the possibility that such a design may extend to their marooning themselves on the motorway island, by the same token it is not inconceivable that the discovery of the flyers (and subsequent retrieval of digital information) might plant the idea of an island visit when no such plan had previously been on the cards. Either way, as at the time of writing there remain no correspondents who have got in touch; I can only conclude that either no one has since ventured along the same route or no one has felt sufficiently interested to want to find out more.[9]

---

[9] A follow-on opportunity in this regard is presented by the open-access nature of this journal article and the accompanying audio–visual content accessible via the social media/video sharing platform YouTube (see [44]). Use

Figure 20. "Concrete Island" project flyer.

It would seem, then, that the island, and the islandness I accrued over the day and night I spent there, are destined to remain mine and mine alone. Perhaps this is as it should be. After all, not only is it illegal to trespass on motorways it is also, self evidently, extremely dangerous. Needless to say this is by no means the kind of project that could be run by a university's research ethics committee for approval. The rationale for this undertaking can hopefully be evinced from the analysis and discussion presented throughout this paper. As a non-place—an unequivocally *negative space*—it is not the topography and landscape of the island that is of particular import (its constitutive placelessness and absence of human dwelling is for very good reason). Rather, it is all that the island *opens up to* as an inverted space of abstraction that is of significance: the negation of the negation of place; the non-place as a launch pad of the imagination; the flow and deterritorialisation of the affective body; a creatively "flirtatious" [35] surfing of the *utopic*; a deep mapping of depthless space. As a negative space, the island, and the islandness it helps foster, is all of these things: a reverse transparency on which to map the meshworks, rhythms and delirious mobilities of late modernity.

Conflicts of Interest

The author declares no conflict of interest.

References

1. J. G. Ballard. *Concrete Island.* London: Fourth Estate, 2014. Original published 1974.
2. Henri Lefebvre. *Rhythmanalysis: Space, Time and Everyday Life*. London: Continuum, 2004.

---

of YouTube comments as part of follow-up analyses in an earlier video mapping/spatial anthropology project called "The Bulger Case: A Spatial Story" provided an additional layer of contextual discussion relating to the issues raised by the video—see Roberts [63]. Time will tell whether similar opportunities present themselves in relation to the Concrete Island project.

3. Tim Edensor. "Introduction: Thinking about Rhythm and Space." In *Geographies of Rhythm: Nature, Place, Mobilities and Bodies*. Edited by Tim Edensor. Farnham: Ashgate, 2010, pp. 1–18.
4. Gaston Bachelard. *The Dialectic of Duration*. Manchester: Clinamen Press, 2000.
5. Les Roberts. "The Violence of Non Places." In *Tourism and Violence*. Edited by Hazel Andrews. Farnham: Ashgate, 2014, pp. 13–31.
6. Meaghan Morris. "At Henry Parkes Motel." *Cultural Studies* 2 (1988): 1–47.
7. Michel de Certeau. *The Practice of Everyday Life*. London: University of California Press, 1984.
8. Les Roberts. "Utopic Horizons: Cinematic Geographies of Travel and Migration." PhD thesis, Middlesex University, 6 July 2005. Available online: www.liminoids.com/Les Roberts _PhD_ 2005.pdf (accessed on 30 July 2015).
9. Les Roberts. "'Welcome to Dreamland': From Place to Non-place and Back Again in Pawel Pawlikowski's Last Resort." *New Cinemas: Journal of Contemporary Film* 1 (2002): 78–90.
10. Hazel Andrews, and Les Roberts. "Introduction: Re-mapping Liminality." In *Liminal Landscapes: Travel, Experience and Spaces In-between*. Edited by Hazel Andrews and Les Roberts. Abingdon: Routledge, 2012, pp. 1–17.
11. Marc Augé. *Oblivion*. Minneapolis: University of Minnesota Press, 2004.
12. Hazel Andrews, and Les Roberts, eds. *Liminal Landscapes: Travel, Experience and Spaces In-between*. Abingdon: Routledge, 2012.
13. Hazel Andrews, and Les Roberts. "Liminality." In *International Encyclopedia of the Social and Behavioral Science*. Edited by Neil J. Smelser and Paul B. Baltes. Oxford: Elsevier, 2015, pp. 131–37.
14. Bjørn Thomassen. *Liminality and the Modern: Living Through the In-Between*. Farnham: Ashgate, 2014.
15. Marc Augé. *Non-Places: Introduction to an Anthropology of Supermodernity*. London: Verso, 1995.
16. Mimi Sheller. "Automotive Emotions: Feeling the Car." *Theory, Culture and Society* 21 (2004): 221–42.
17. Iain Borden. *Drive: Journeys through Film, Cities and Landscapes*. London: Reaktion, 2013.
18. Nigel Thrift. "Driving in the City." *Theory, Culture and Society* 21 (2004): 41–59.
19. Peter Merriman. "Driving Places: Marc Augé, Non-places, and the Geographies of England's M1 Motorway." *Theory, Culture and Society* 21 (2004): 145–67.
20. Peter Merriman. *Driving Spaces: A Cultural-Historical Geography of England's M1 Motorway*. Oxford: Blackwell, 2007.
21. Edward Dimendberg. *Film Noir and the Spaces of Modernity*. London: Harvard University Press, 2004.
22. Iain Sinclair. *Lights Out for the Territory*. London: Penguin, 2003.
23. Henri Lefebvre. *The Production of Space*. Oxford: Blackwell, 1991.
24. Iain Sinclair. *London Orbital: A Walk around the M25*. London: Penguin, 2003.
25. Nick Papadimitriou. *Scarp*. London: Hodder & Stoughton, 2012.

26. Margaret Morse. "An Ontology of Everyday Distraction." In *Logics of Television: Essays in Cultural Criticism*. Edited by Patricia Mellencamp. Bloomington: Indiana University Press, 1990, pp. 193–219.
27. Simon Ward. "'Danger Zones': The British 'Road Movie' and the Liminal Landscape." In *Liminal Landscapes: Travel, Experience and Spaces In-between*. Edited by Hazel Andrews and Les Roberts. Abingdon: Routledge, 2012, pp. 185–99.
28. Michael Bracewell. "Fade to Grey: Motorways and Monotony." In *Autopia: Cars and Culture*. Edited by Peter Wollen and Joe Kerr. London: Reaktion, 2002, pp. 288–92.
29. Tim Edensor. "M6—Junction 19-16: Defamiliarizing the Mundane Roadscape." *Space and Culture* 6 (2003): 151–68.
30. Peter Osborne. "Non-places and the Spaces of Art." *The Journal of Architecture* 6 (2001): 183–94.
31. Louis Marin. *Utopics: The Semiological Play of Textual Spaces*. Atlantic Highlands: Humanities Press International, 1984.
32. The Guardian. "Deadly fantasist." Available online: http://www.theguardian.com/theguardian/2001/jan/13/weekend7.weekend1 (accessed on 18 September 2015).
33. Les Roberts. *Film, Mobility and Urban Space: A Cinematic Geography of Liverpool*. Liverpool: Liverpool University Press, 2012.
34. Chris Lukinbeal, and Leo Zonn, eds. "Cinematic Geographies."*GeoJournal* 59 (2004): 247–337.
35. David Crouch. *Flirting with Space: Journeys and Creativity*. Farnham: Ashgate, 2010.
36. Tim Edensor. "Commuter: Mobility, Rhythm and Commuting." In *Geographies of Mobilities: Practices, Spaces, Subjects*. Edited by Tim Cresswell and Peter Merriman. Farnham: Ashgate, 2012, pp. 189–203.
37. Benjamin Fraser. *Toward an Urban Cultural Studies: Henri Lefebvre and the Humanities*. New York: Palgrave, 2015.
38. Marc Augé. "Roundabouts/Yellow Lines." In *Autopia: Cars and Culture*. Edited by Peter Wollen and Joe Kerr. London: Reaktion, 2002, pp. 293–95.
39. Paul Virilio. *Negative Horizon: An Essay in Dromoscopy*. London: Continuum, 2005.
40. Giuliana Bruno. *Streetwalking on a Ruined Map: Cultural Theory and the City Films of Elvira Notari*. Princeton: Princeton University Press, 1993.
41. Michael Bull. "Soundscapes of the Car: A Critical Ethnography of Automobile Habitation." In *Car Cultures*. Edited by Daniel Miller. Oxford: Berg, 2001, pp. 185–202.
42. "Concrete Island sound clip 1." Available online: www.liminoids.com/inthefield/ConcreteIsland_soundclip1.mp3 (accessed on 3 August 2015).
43. "Concrete Island sound clip 2." Available online:www.liminoids.com/inthefield/ConcreteIsland_soundclip2.mp3 (accessed on 3 August 2015).
44. "Concrete Island video." Available online: https://youtu.be/LP5tQLDc_Vo (accessed on 3 August 2015).
45. Department of the Environment, and North Western Road Construction Unit. *M53 Motorway*. Chester: Motorway Archive Trust, Cheshire Records Office, n.d.
46. William James McCoubrey. *The Motorway Achievement, Volume 3: Building the Network*. London: Thomas Telford, 2008.

47. Maunsell and Partners. *Report on the Mid-Wirral Road.* Chester: Motorway Archive Trust, Cheshire Records Office, 1966.
48. Phillip Vannini, and Jonathan Taggart. “Doing Islandness: A Non-Representational Approach to an Island’s Sense of Place.” *Cultural Geographies* 20 (2012): 225–42.
49. Cheshire Records Office. “Cheshire Archives and Local Studies.” Available online: http://archives.cheshire.gov.uk/ (accessed on 17 July 2015).
50. Pathetic Motorways. “M53 Mid-Wirral Motorway.” Available online: http://pathetic.org.uk/current/m53/ (accessed 17 July 2015).
51. The Society for All British and Irish Road Enthusiasts. Available online: www.sabre-roads.org.uk/ (accessed on 17 July 2015).
52. Wiki Wirral.co.uk. “The M53 (+M531)—The History Of Wirrals Motorway.” Available online: www.wikiwirral.co.uk/forums/ubbthreads.php/topics/101185/The_M53_M531_The_History_Of_Wi.html (accessed on 17 July 2015).
53. Les Roberts, and Hazel Andrews. “(Un)Doing Tourism Anthropology: Outline of a Field of Practice.” *Journal of Tourism Challenges and Trends* 6 (2014): 13–38.
54. Tim Ingold. *The Perception of the Environment: Essays in Livelihood: Dwelling and Skill.* London: Routledge, 2000.
55. Les Roberts. “Mapping Cultures—A Spatial Anthropology.” In *Mapping Cultures: Place, Practice, Performance*. Edited by Les Roberts. Basingstoke: Palgrave, 2015, pp. 1–25.
56. Simon Gottschalk, and Marko Salvaggio. “Stuck Inside of Mobile: Ethnography in Non-Places.” *Journal of Contemporary Ethnography* 44 (2015): 3–33.
57. Victor Turner. *The Ritual Process: Structure and Anti-Structure*. New York: Cornell University Press, 1969.
58. Michael Jackson. *Minima Ethnographica: Intersubjectivity and the Anthropological Project.* Chicago: University of Chicago Press, 1998.
59. Michael Jackson. *Paths toward Clearing: Radical Empiricism and Ethnographic Inquiry.* Bloomington: Indiana University Press, 1989.
60. Michael Jackson. “Introduction. Phenomenology, Radical Empiricism, and Anthropological Critique.” In *Things as They Are: New Directions in Phenomenological Anthropology*. Edited by Michal Jackson. Bloomington: Indiana University Press, 1996, pp. 1–50.
61. Michael Jackson. *Existential Anthropology: Events, Exigencies and Effects.* Oxford: Berghahn, 2005.
62. Dimitra Gefou-Madianou. “Ethnography in Motion: Shifting Fields on Airport Grounds.” In *Ethnographic Practice in the Present*. Edited by Marit Melhuus, Jon P. Mitchell and Helena Wulff. Oxford: Berghahn, 2009, pp. 152–68.
63. Les Roberts. “The Bulger Case: A Spatial Story.” *The Cartographic Journal* 51 (2014): 141–51.

Archaeological Excavation and Deep Mapping in Historic Rural Communities

Carenza Lewis

Abstract: This paper reviews the results of more than a hundred small archaeological "test pit" excavations carried out in 2013 within four rural communities in eastern England. Each excavation used standardized protocols in a different location within the host village, with the finds dated and mapped to create a series of maps spanning more than 3500 years, in order to advance understanding of the spatial development of settlements and landscapes over time. The excavations were all carried out by local volunteers working physically within their own communities, supported and advised by professional archaeologists, with most test pits sited in volunteers' own gardens or those of their friends, family or neighbors. Site-by-site, the results provided glimpses of the use made by humans of each of the excavated sites spanning prehistory to the present day; while in aggregate the mapped data show how settlement and land-use developed and changed over time. Feedback from participants also demonstrates the diverse positive impacts the project had on individuals and communities. The results are presented and reviewed here in order to highlight the contribution archaeological test pit excavation can make to deep mapping, and the contribution that deep mapping can make to rural communities.

Reprinted from *Humanities*. Cite as: Lewis, C. Archaeological Excavation and Deep Mapping in Historic Rural Communities. *Humanities* **2015**, *4*, 393–417.

1. Introduction

Deep mapping, as epitomised by William Least Heat-Moon in PrairyErth [1], encompasses a range of approaches—cartographic, geographical, historical, literary, philosophical, scientific, anthropological, sociological and theological—to weave multiple strands of evidence, observation, impression and memory into a distillation of place which is part local history, part biography, part travelogue [2]. Heat-Moon's work occupies just one point within a vast and wide-ranging multi-disciplinary profusion of diachronically inflected, socially informed studies of place, whose various practitioners might or might not identify themselves variably as local historians, writers, performers, film-makers, philosophers, farmers or householders [3–9]. While travel writing, folklore studies and local history have often lain at (or beyond) the margins of academic scholarship, research espousing many of the tenets of deep mapping can be found within disciplines such as historical geography [10], psychogeography [11], landscape archaeology [12] and the newer field of spatial humanities [13]. Modern technology is catalysing innovation, with digital geographical information systems (GIS) now enabling scholars to combine "bottom-up" data and "top-down" technology in new ways [14,15] to create spatially articulated, scalable representations of space which can encompass a range of natural, built, tangible and intangible heritage [16].

Exploring time depth is often an explicit aim of deep mapping, reflecting Heat-Moon who described PrairyErth as a "vertical journey" through time and referred often to archaeology when presenting his material and discussing his approaches [17]. Archaeology itself is a discipline which aims is to understand the past through the study of its physical remains and has long embraced spatial thinking: many archaeologists create maps which do include evidence found beneath the ground, created at scales ranging from the individual site to entire continents [18]. Deep mapping and archaeology might thus seem to have compatible aims that would encourage effective interdisciplinary working in pursuit of common goals.

However, perhaps surprisingly, this has not proved to be the case [19]. Despite frequent references to verticality, little of the deep mapping literature involves actual penetration below the physical surface of the explored terrains: Heat-Moon uses archaeological excavation as a metaphor rather than a practised technique, while recent deep maps such as Ethington and Toyosawa's palimpsest of Los Angeles reveals a rich history, but no data from below the surface [20]. Within archaeology, antiquarian chorographic approaches to holistically representing places in time have fallen from favour, while more recent attempts at revival and reinvention of this tradition have been acknowledged as too protean to gain widespread academic traction [21]. GIS has been used by archaeologists since the 1980s to map and interrogate diverse datasets [22,23], but such studies, while displaying increasing technical virtuosity, can be overly absolutist and lose connection with humans, past and present [24,25]. Conversely, phenomenological studies within archaeology which since the 1990s have aimed to foreground the past lived human experience [26] have generated inferences which appear overly subjective and difficult to substantiate. Crowdsourced projects [27] represent a different approach again to involving contemporary publics in generating dynamic, expanding, continuously evolving place maps with time depth but such projects do not always prioritise reflexive scholarly analysis.

Such difficulties in creating meaningful links between the "soft" people-centred inputs and "hard" archaeologically-visible outputs of human societies, in the past and the present are familiar to those involved in public archaeology [28], but are not always easily resolved, especially if volunteers, researchers and communicators have incompatible or conflicting interests or priorities. This paper reviews recent research in which new archaeological excavations were carried out by residents of rural villages within the spaces they currently inhabit in order to make new discoveries about the past history of those places. The excavations were carried out in collaboration with university researchers who provided technical and methodological advice and on-site supervision as well as specialist evaluation of the results which were reported on jointly with community groups. These projects about communities, in communities, for communities, with communities and by communities generated data from more than 100 individual sites within four villages which could be aggregated to create new maps showing how spaces and places occupied by today's communities were used in the past spanning thousands of years. This activity contributes to academic research as well as local knowledge, while also connecting volunteers with the pasts of the spaces they live in today and building social capital within communities for the future. The completion of a number of similar projects shows that their outcomes can be achieved widely to significant public benefit.

2. The Cambridge Community Heritage (CCH) Project

The four projects presented here formed part of the "Cambridge Community Heritage" (CCH) programme [29] which was funded in 2012–2013 under the Arts and Humanities Research Council (AHRC) Connected Communities theme's *Research for Community Heritage* (R4CH) call [30] jointly with the Heritage Lottery Fund (HLF) *All Our Stories* [31] fund. R4CH was intended to help community groups explore their heritage by giving them access to resources and expertise that exists within universities, and to create new opportunities for academics to work in a community context. Community groups chose for themselves the aspect of their heritage they wished to explore, and were the main drivers in deciding what approaches should be used to achieve this. In 2012 CCH issued an open call to community groups in eastern England (Norfolk, Suffolk, Cambridgeshire, Essex, Bedfordshire and Hertfordshire) for expressions of interest in heritage-related community project of any type. Submitted ideas ranged extremely diversely across tangible and intangible heritage, proposed by groups representing communities of place, occupation, interest and identity including local historical societies, football clubs, church groups, traveller communities, schools, women's groups and military regiments [32].

Out of a total of 23 community groups who were successful in securing funding within CCH, four were local history societies, based in the southern English villages of Meldreth, West Wickham, Toft and Shillington (Figure 1), who wished to find out more about their community's history through excavating small archaeological "test pits". The test pit approach, which involves the excavation of standard-sized metre-square in numerous different places across a site such as a historic settlement, can be used to recover buried archaeological data from largely built-up environments (such as today's villages, hamlets and towns), within which open-area excavation of larger trenches is impossible [33]. Dating and mapping the distribution of finds recovered from the test pits, in particular worked flint and pottery sherds which are datable and found in large numbers, allows the changing spatial pattern of activity over time to be reconstructed [34,35]. Within the field of historic settlement studies, work in these currently occupied rural settlements (CORS) is an important priority as it allows the development of non-deserted settlements to be included in research which otherwise focusses primarily on deserted sites which are more easily accessible for archaeological excavation [36–39].

The CCH excavations in Meldreth, Shillington, Toft and West Wickham involved hundreds of local residents. Participation was open to everyone who lived in the communities, and to friends and family members resident elsewhere, and could include excavating and/or offering a site for excavation and/or helping with project planning and organisation. Within each excavation team, opportunities were present for both able-bodied and less able people of all ages (participants ranged in age from a few months to over eighty years) to take part in a wide range of activities including digging into the ground, searching through excavated spoil, finds washing and maintaining written records (the task of preparing refreshments was not a formal part of the process but drew in many other helpers). More than 100 separate excavations were completed, from which worked flint and pottery sherds were identified, dated and located to produce a series of maps showing which plots, streets and neighbourhoods within each of the four communities had produced finds from which historic periods. This information was contextualised and assessed in a technical report drafted for each settlement [40–43] summarised for publication in *Medieval Settlement Research* (e.g., [44]) and are contributing to ongoing academic research into the development of settlement, landscape and demography in southern England [33,45].

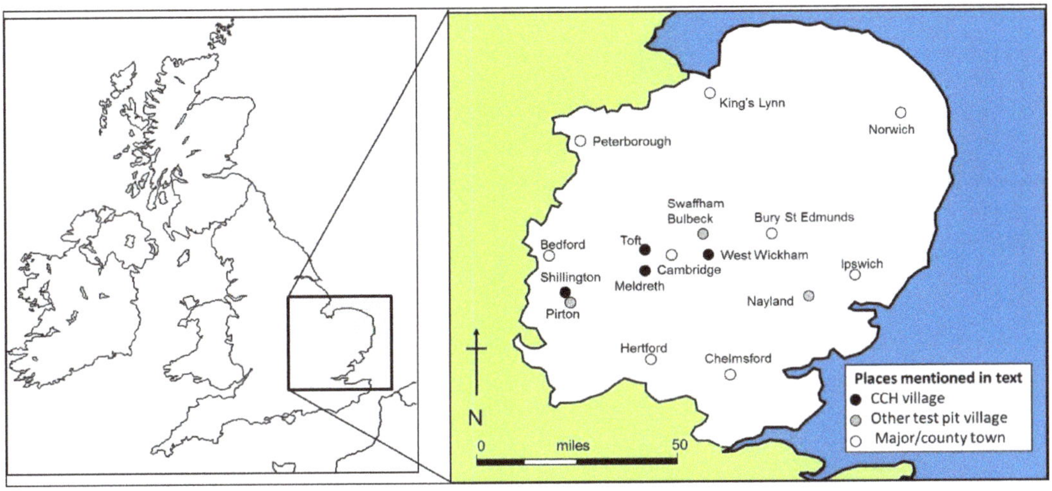

Figure 1. Map of the British Isles showing places mentioned in the text.

2.1. Meldreth

The village of Meldreth is situated in south Cambridgeshire, England, 15 km southwest of Cambridge, centred on TL 37610 45812 (Figure 1). The parish lies in the valley of the River Cam or Rhee, which marks the northern parish boundary. The River Mel rises immediately south of the village at Melbourn Bury and runs through the village along the back of properties on High Street, before joining the Rhee north of the parish church. The parish of Meldreth lies on fairly flat and low-lying Cretaceous chalk bedrock between 15 m and 40 m above OD. The surrounding landscape is presently composed of gently rolling open arable farmland with drainage ditches and small streams and fragmented hedgerows forming field boundaries.

The existing village of Meldreth (Figure 2) is broadly linear in layout on an N-S orientation immediately west of the River Mel, forming an almost continuous polyfocal settlement more than 2 km long running between the neighbouring villages of Shepreth (to the north) and Melbourn (to the south). Analysis of the first edition of the 6-inch to 1 mile Ordnance Survey map shows that in the latter part of 19th century settlement at Meldreth was more dispersed than it is today and less continuous, divided into several discrete elements. The largest of these along High Street (Figure 2(Aa)) (immediately west of the River Mel), comprising a north-south-orientated nucleated planned linear row approximately 400 m long. This was separated by more than 200 m from three small clusters of settlement to its south (around the railway station (Figure 2(Ab)), its west (Chiswick End (Figure 2(Ac))) and its north (Manor Farm (Figure 2(Ad))). Approximately 200 m north-east of Manor Farm there was a smaller, less regular east-west orientated row running past the church (Figure 2(Ae)) and another even further to the north-east at North End (Figure 2(Af)), arranged around a small green. An isolated farm lay south of the village beyond the railway line (Figure 2(Ag)).

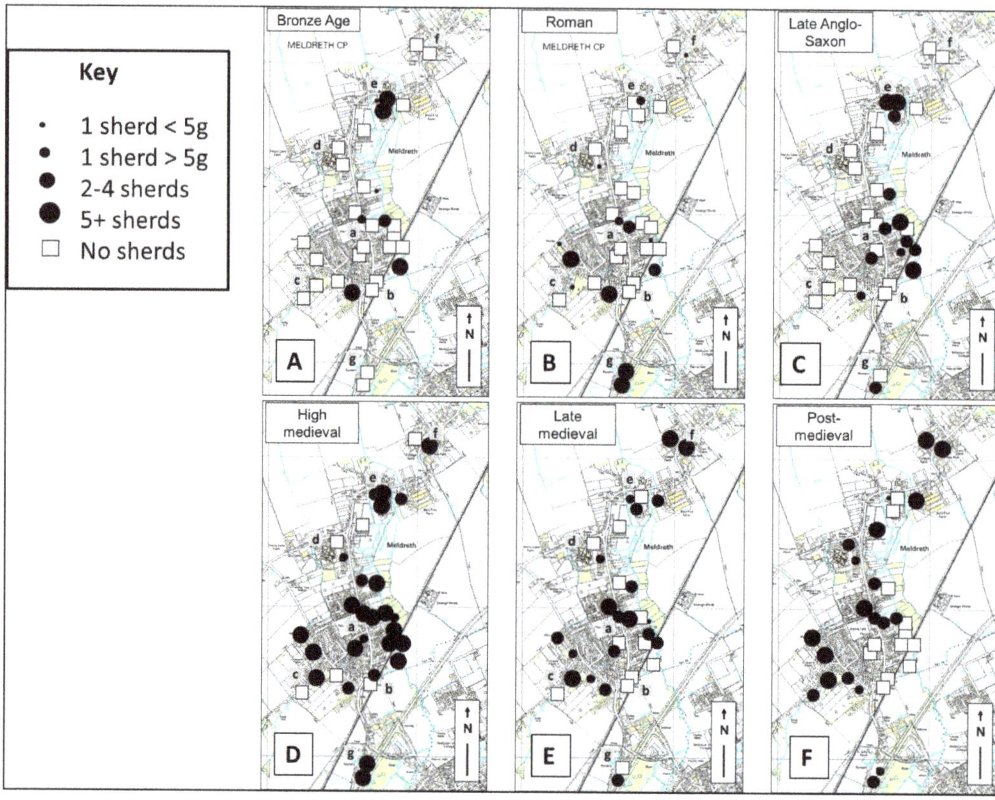

Figure 2. Pottery from test pits excavated in Meldreth. (A) Bronze Age; (B) Roman; (C) Late Anglo-Saxon; (D) High medieval; (E) Late medieval; and (F) Post-medieval.

The 2013 test pit excavations were coordinated by Meldreth Local History Group [46] and involved more than 300 local residents in excavating 32 test pits ([47]; [48], pp. 120–21) as part of an extended suite of community-centred activities. These included a village hall lecture by the supervising university archaeologist delivered some months before the excavations, explaining their aims and speculating on the site-specific potential of the Meldreth excavations. The excavations themselves were each carried out over two days during one of three weekends spaced at intervals of approximately one month in summer 2013. Each of these weekends began with a short briefing explaining the excavation process (which had to be closely followed in order to ensure data validity) to participating volunteers (local residents along with families, friends and neighbours), who then dispersed to their gardens, or those of the friends or neighbours, to begin the excavation. Each site was visited frequently during the two days by a professional archaeologist from the university in order to ensure required procedures were being followed, provide advice and talk to volunteers about their finds and more generally about aspects of the excavations or their property which interested them. One resident filmed the talks and as much of the excavations as possible. At the end of the second, final day of each digging weekend, each team brought the finds from their test pit and the records they had kept to a central venue where tea and cake was provided and volunteers could view each other's finds. A representative of each team then provided a short verbal summary of their excavation to the assembled group, chaired by the supervising archaeologist who also provided an assessment of the emerging "bigger picture" generated by the new finds in their final summary. In the months following the excavations, the results were formally analysed and contextualised and a technical report prepared by the university CCH team [41], an exhibition was held in the village hall in which finds were displayed along with a range of other local historical research carried out by local volunteers and the final results were presented to the local community in another talk held the following winter, which included new maps which included the excavated date and showed how the place had changed over time.

Analysis of the excavated finds (Figure 2) showed Bronze Age pottery (c. 1800–800 BC) to have been recovered from several different areas (Figure 2A) in the centre and north of the present village. Such finds are rare from test pits and this quantity from Meldreth suggests unusually intensive or persistent use by humans around three thousand years ago of the part of the landscape covered by the present village, especially around ("e"). This may have involved either settlement or mortuary activity, or possibly both. No evidence for the succeeding Iron Age period (800 BC–43 AD) was found from any of the pits, but this does not necessarily suggest the area was unused at this time, as pottery is relatively scant in settlements of this date. Pottery of Roman date (43–410 AD) (Figure 2B) was found, however, mostly in the centre and south of the present village (Figure 2(Ba–c and Bg)) in a pattern which suggested that settlement by then took the form of a dispersed scatter of small settlements such as farmsteads. Smaller amounts of pottery from test pits to the north suggest that area may then have been in less intensive use, possibly as arable fields. No pottery was found dating to the period between the 5th–9th centuries AD, but this is not unusual as settlements as this time tend to be small and produce little pottery therefore are easily missed by test pits, so this does not necessarily indicate the area was unused by humans at this time. In contrast, however, pottery of late Anglo-Saxon date (mid-9th to late 11th century AD) (Figure 2C) was found widely in two different areas (Figure 2(Ca,Ce)). A concentration from three pits in the north of the present village (Figure 2C,E) suggest the later-documented manorial site of Topcliffe originated at this date, while pottery from seven pits in the centre of the village (Figure 2C, west of "a"), all between the present High Street and the river, suggest the settlement at this time formed a linear row on one side of the street only, with smaller amounts of pottery from pits west of High Street suggesting this side of the street was used as arable fields. This pattern of settlement is significantly different to that in the Roman period, with much less Anglo-Saxon pottery found to the south and south-west of the present village (areas "b", "c" and "g") than in the Roman period, suggesting that use of this area may have changed from dispersed settlement to arable fields.

In the subsequent period, the amount of pottery of high medieval date (12th–14th century AD) (Figure 2D) indicates that in this period the settlement increased in size. Several pits west of the High Street (Figure 2(Da)) produced high medieval pottery, suggesting that new tofts were added in the 12th or 13th century along the west side of the High Street, probably laid out over former arable (explaining and dating the narrow property boundaries still observable on the 19th century maps).

Other new additions to the settlement pattern at this time appear to have included smaller dispersed hamlets at Chiswick (west of the present village) (Figure 2(Dc)), North End (at the far north of the present village) (Figure 2(Df)) and near the present garden centre (Figure 2(Dg)), possibly the site of a separate farm. Pottery dating to the late medieval period (14th–16th centuries) (Figure 2E) was found in fewer pits and in lesser amounts, suggesting that settlement expansion ceased at this time, although it does not appear to contract in size significantly at his time, although the volume of late medieval pottery recovered from two historically documented medieval manorial sites (Topcliffe (Figure 2(De)) and Flambards (Figure 2D south-east of "a") is dramatically less than previously. Overall, comparison with settlements elsewhere suggests that Meldreth was less severely by demographic decline than many settlements in the eastern region ([33], pp. 330–434; [45]). In the subsequent post-medieval period (16th–18th centuries) (Figure 2F), the test pit data suggest that Meldreth stagnated, with the south-eastern part of the settlement (Figure 2F, between "a" and "b"), which had produced large amounts of earlier pottery, particularly badly affected.

2.2. Shillington

Shillington is situated in south Bedfordshire near the border with Hertfordshire, 17 km southeast of Bedford, centred on TL 12562 34625 (Figure 1). The present settlement lies between 50–60 m above OD between the Pegsdon Hills a few kilometres to the south (part of the Chilterns), and the Greensand Ridge in Bedfordshire to the north. The parish lies among the headwaters of the Ouse catchment and its southern boundary follows the course of the Icknield Way over a spur of the Chiltern Hills at Pegsdon. A small brook flows in a northerly direction on the west side of a prominent hill and drains into the River Hitt and ultimately the Great Ouse. The surrounding land is today broadly composed of gently rolling open farmland with drainage ditches, water courses and fragmented hedgerows forming field boundaries.

Settlement in Shillingont parish today (Figure 3), which includes the formerly separate parish of Higham Gobion and the village of Pegsdon, is moderately dispersed, extending over more than 2 km along a succession of lanes which loop around to form two large polygons. A largely continuous stretch of housing extends between Woodmer End (Figure 3(Aa)) through Hillfoot End (Figure 3(Ab)) to Marquis Hill (Figure 3(Ec)) and east and south of the 14th century parish church of All Saints (located on the summit of the prominent chalk hill) (Figure 3(Ad)), affording it clear views of the surrounding landscape and rendering it visible from some distance. Settlement elsewhere in Shillington today is much more dispersed, arranged in several "Ends" including Apsely End (Figure 3(Ae)), Hanscombe End (Figure 3(Af)), Upton End (Figure 3(Ag)) and Bury End (Figure 3(Ah)). Discrete farms sited around the settlement include Hanscombe End Farm, Moorhen Farm, Northley Farm, Lordship Farm, Upton End Farm and Clawders Hill Farm.

The 19th century settlement, as depicted on the first edition Ordnance Survey 6-inch to one mile scale map, sprawled across an equally large area but contained fewer houses and retained a very much more dispersed character. The greatest concentration of housing was along Church Street (Figure 3A south-east of "d") running east from the church), which was flanked by a nucleated double row of housing (although several plots north of this street were devoid of buildings), extending into the lane leading north towards Hillfoot End (Figure 3(Ab)). The church was then on the very edge of the settlement, with no houses then present to its north or west. To its south, housing was much more intermittent along the north side of High Road, with an intermittent succession of small properties forming an interrupted row extending north-east as far as Marquis Hill (Figure 3(Ac)) where the settlement petered out. Hillfoot End (Figure 3(Ab)) in the 19th century was an entirely separate hamlet comprising around a dozen or so cottages mostly south-east of a tiny triangular green where three lanes meet. Hanscombe End (Figure 3(Af)) was extremely dispersed with a handful of properties of varying size arranged along a winding lane. Woodmer End (Figure 3(Aa)) comprised perhaps 20 properties along a single lane arranged as an interrupted row at the south end and a more compact double row to the north, where it merged with the similar but smaller hamlet of Bury End (Figure 3(Ah)). Upton End (Figure 3(Ag)) comprised just 4–5 larger properties either side of the road towards Marquis Hill (Figure 3(Ac)), where there is very little settlement at all. Northley Farm and Shillington Bury Farm were isolated sites with no near neighbours.

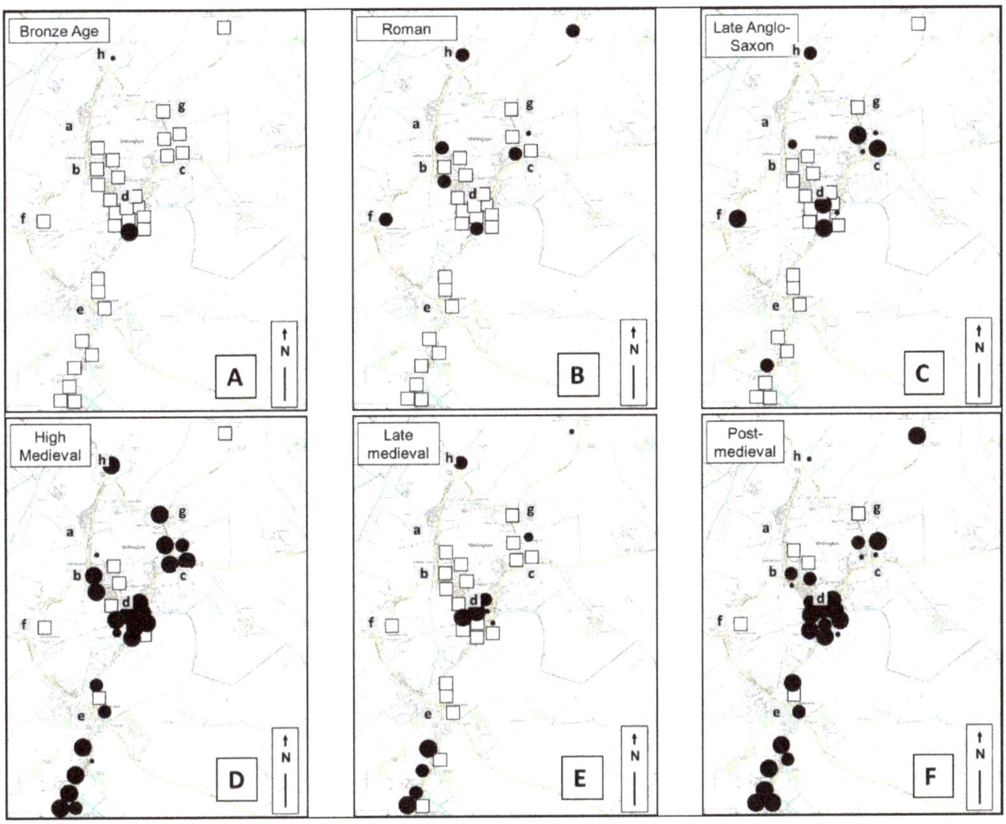

Figure 3. Pottery from test pits excavated in Shillington. (A) Bronze Age; (B) Roman; (C) Late Anglo-Saxon; (D) High medieval; (E) Late medieval; and (F) Post-medieval.

The excavations were coordinated by Shillington Local History Society [49] and involved more than 300 residents of the village of Shillington and its local area in completing 23 test pits throughout the present village over a single two-day weekend in summer 2013. This began with refreshments and a short briefing after which volunteers dispersed to their chosen or allocated sites to begin the excavation. As at Meldreth, each site was visited frequently by supervising archaeologists and the second, final day ended with refreshments and a finds-viewing plenary in which each team reported on their finds. The finds were subsequently analysed and a report prepared [42] which mapped the finds and showed changes in the spatial disposition of human use of the landscape and settlement over time.

The 2013 excavations at Shillington produced Bronze Age pottery from two different locations in the north and centre of the area covered by the present parish (Figure 3(Ad,Ah)), suggesting a scattered pattern of land-use. Most notably, a test pit (Figure 3A, south of "d"), near the small brook running west of the prominent chalk hill produced sufficient material to indicate intensive activity such as settlement or burial. Romano-British pottery came from seven different sites (Figure 3B), two of them (Figure 3(Bc) and north of "g") away from the brookside area hinting at activity in this period extending beyond the lower lying stream-side zone. No evidence was found for any activity dating to the period between the 5th–9th centuries AD, but late Anglo-Saxon pottery (9th–11th century date) was found in two distinct concentrations (Figure 3(Cc,Cd)) sufficient to suggest the presence of small nucleated settlements with three other locations (Figure 3(Ce,Cf and Ch)) yielding sufficient pottery to hint at the possibility of outlying dispersed hamlets or farms present at this time. The high medieval period saw all these settlements grow, creating two apparently sizeable, probably nucleated, settlements south of the church and at Marquis Hill (Figure 3(Dd,Dc)) and a probably more dispersed attenuated row settlement up to 750 m long at Aspley End (Figure 3D, south of "e"). At the same time, habitative activity appears to commence at Hillfoot End and Upton End (Figure 3(Db,Dg)), indicating a pattern of mixed dispersed and nucleated settlement. This growth in settlement size and number was thrown into reverse in the late medieval period, with the mapped pottery data (Figure 3E) suggesting Shillington was particularly badly affected in this period of widespread demographic and settlement contraction compared to many settlements in the eastern region. Habitation at Marquis Hill, Upton End, Hillfoot End (Figure 3(Ec,Eg and Eb)) and the north of Aspley End (Figure 3E, north of "e") appears to have ceased almost completely. In the post-medieval period, however, the test pit data indicates that Shillington gradually recovered (Figure 3F) , with the area around the church (Figure 3(Fd)) and most of the high medieval dispersed settlements reoccupied, although some of the medieval "ends" remained uninhabited for longer, especially in the north of the parish (Figure 3(Fb,Fh and Fg)).

2.3. *Toft*

The village of Toft lies in the county of Cambridgeshire, 9 km SWW of Cambridge and just 10 km north of Meldreth centred on TL 3596 5600 (Figure 1). Toft is one of several parishes lying on the northern bank of the Bourn Brook, which rises a few miles west of Toft and joins the River Cam just south of Grantchester. Toft parish lies on a south-facing slope between 25 m and 40 m OD on Cretaceous sedimentary Gault mudstone bedrock which is capped in the northern part of the parish by superficial Quaternary deposits of sands and gravels. The surrounding landscape is today broadly composed of gently rolling open arable farmland with drainage ditches and small streams and fragmented hedgerows forming field boundaries.

The excavations in 2013 focussed on the south of the present village of Toft, which comprises a nucleated settlement arranged either side of two main approximately north-south-orientated streets, High Street (Figure 4(Aa)) and School Lane (Figure 4(Ab)), and another row running perpendicular to these along Comberton Road running gently upslope (Figure 4(Ac)). The medieval parish church (Figure 4(Ad)) is now largely isolated and lies adjacent to fields south east of the present village.

Figure 4. Pottery from test pits excavated in Toft. (A) Roman; (B) Early Anglo-Saxon; (C) Late Anglo-Saxon; (D) High medieval; (E) Late medieval; and (F) Post-medieval.

The 19th century Ordnance Survey maps shows the settlement then to have been much smaller and more dispersed than it is today, arranged loosely around a square grid of lanes between Comberton Road and the Bourn Brook. The most compact part of the settlement then was arranged as a linear row along the Comberton Road near its junction with Church Road (Figure 4(Ae)). Settlement along the northern end of the High Street (Figure 4A, north of "a") was much more intermittent than today, constituting little more than a Methodist chapel and an inn. This is separated by some 150 m from a small north-west-south-east orientated single row of farms and cottages along a lane now called Brookside (then called Water Row) (Figure 4(Af)), which appears then to have constituted a separate hamlet. The dispersed character of the settlement pattern in this area is further emphasised by the presence of just a couple of cottages along School Lane, leaving the church even more isolated than it is today, with only the Rectory and a small terrace of three cottages for company.

The CCH test pit excavations at Toft were instigated by Toft Historical Society [50]. As in Meldreth, the project included a range of linked activities which in Toft included a village archaeological survey exploring earthwork remains in an area of deserted settlement, the excavations during which more than 600 residents of the village and the local area took part in excavating or visiting excavations in 16 different locations throughout the present village, the making of a film, preparation of a written technical report [43], a post-excavation winter talk on the results and a public exhibition when finds and the film were shown.

Analysis of the results showed that the area encompassed by the excavations was intermittently and lightly used by humans in the prehistoric period, with possible indications of a small settlement of Neolithic date beside the brook, beyond the south-eastern limits of the present village. No later prehistoric material was found, but the distribution of pottery of Roman date (Figure 4A) suggests a settlement of some substance to then have been present then in the same brook-side area (Figure 4(Ag)), with smaller numbers of sherds found in some of the pits to the north and north-west (Figure 4A near "a" and "b") suggesting the present village may overlie an area in use for arable cultivation during the Roman period. The area around the church (Figure 4(Ad)) also produced small amounts of sherds of Roman date and may possibly indicate contemporary settlement nearby. One small sherd of pottery dating to the 5th to 9th century AD (Figure 4B, west of "g") is a relatively unusual find which hints at the possibility that the Roman stream-side settlement continued into the post-Roman period. This same area (immediately north of the brook and about 200 m south-west of the church) also produced a considerable amount of late Anglo-Saxon pottery (Figure 4(Cg)), suggesting a small settlement, possibly a nucleated village, was by then present in this same location. Whether this represents continuation from the early Anglo-Saxon period or a new foundation cannot be ascertained from the known data—either scenario must currently be considered to be possible. Smaller amounts of pottery found test pits north-west of the brook-side area (Figure 4C, west of "f" and "a") suggest that the arable fields of the Anglo-Saxon settlement lay in this direction, continuing the landscape use pattern of the Roman period.

In the high medieval period (Figure 4D) the settlement appears to have expanded markedly, with many more pits producing pottery of this date, notably in the area north-west of the Anglo-Saxon core (Figure 4(Df)) and south of the Bourn Brook (Figure 4(Dh)). The latter produced no material of any earlier date whatsoever and appears to be newly used for habitation at this time of known demographic growth in England. In the succeeding period (Figure 4E), by contrast, Toft experienced severe contraction, with the stream-side sites (Figure 4(Eg,Eh)) entirely abandoned in the late medieval period. When the settlement began to recover in the post medieval period (Figure 4F), apparently rather falteringly, its focus appears to have shifted north and west (Figure 4F, around "a", "f" and possibly "c"), mirroring the extent of the present settlement, with the areas beside the stream settlement remaining permanently deserted. The demographic crisis of the 14th century can thus be seen to have terminated the history of settlement immediately north of the brook ("g") spanning more than a millennium, as well that of the much more recently colonised area south of the brook ("h").

2.4. West Wickham

West Wickham lies 19 km south-east of Cambridge near the county boundary between Cambridgeshire and Suffolk (Figure 1). The parish is mainly on tertiary chalk underlying an extensive drift cover of glacial clays with small areas of brickearths, with the present village situated on a gradually undulating ridge. There are no streams in the village, however a line of springs present along a gravel ridge running parallel and c. 100 m west of the High Street feed a number of ponds which are still extant to the rear of many of the houses. The present village lies near the centre of the parish and is mostly arranged along a linear High Street (Figure 5(Aa)) running north-east from the parish church of St. Mary (Figure 5(Ab)) for about 1.2 km. Streetly End (Figure 5(Ac)) is a separate hamlet of around a dozen houses located about 0.8 km south of West Wickham.

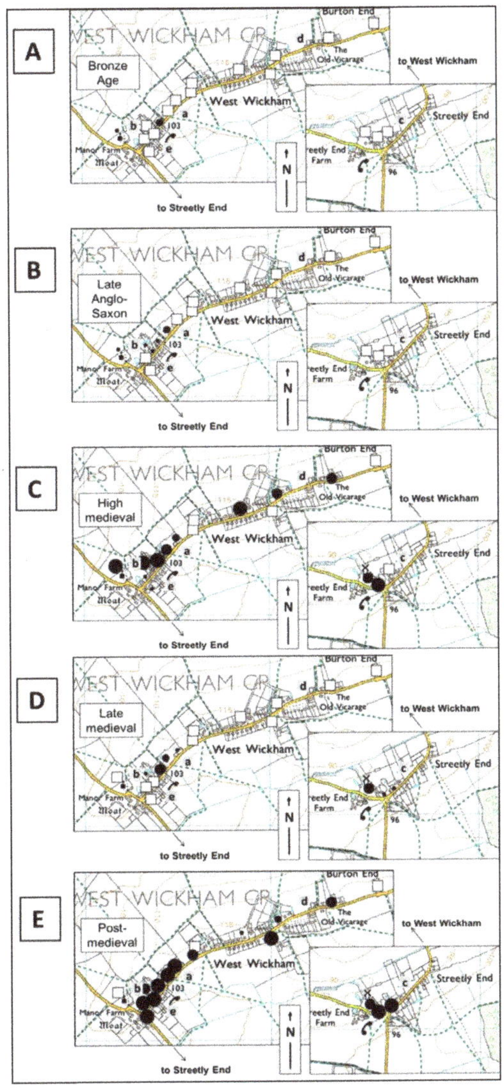

Figure 5. Pottery from test pits excavated in West Wickham. (A) Bronze Age; (B) Late Anglo-Saxon; (C) High medieval; (D) Late medieval; and (E) Post-medieval.

The 19th century Ordnance Survey map shows West Wickham at this time to be formed of two discrete elements separated by more than 200 m. The compact linear settlement near the church favours the north (upslope) side of the High Street and is composed of mostly short plots, while Burton End (Figure 5(Ad)) has a more dispersed character, arranged as an interrupted row of fewer than ten residences most separated by fields. Streetly End appears much as it does today, although the pattern of plot boundaries suggests the High Street there has cut diagonally across an earlier more regular planned block of settlement.

The CCH excavations at West Wickham were organised by the West Wickham District Local History Group [51] and involved more than 70 local people in the excavation of 18 archaeological test pits over a single weekend. These followed the standard procedure of initial technical briefing accompanied by refreshments before the excavations, frequent visits by professional archaeologists during the weekend, a final get-together with refreshments, finds-viewings and reports from each team, followed by formal finds analysis and technical reporting [40] and a winter lecture in the village on the results.

The excavations suggested that the site of West Wickham was only lightly used by humans before the tenth century AD, although a slight concentration of finds of worked flint at the west end of West Wickham (Figure 5(Ae)) was present in contexts which suggest more intensive Bronze Age activity (c. 1800–800 BC), possibly settlement, in the in the area south-east of the (much later) church. Somewhat surprisingly, given finds elsewhere in the parish, no evidence of Roman date was found in any of the test pits within the present village. Material of early or middle Anglo-Saxon date was also absent, but pottery of late Anglo-Saxon date found near the parish church of St. Mary (Figure 5B either side of "b"), suggests that the present settlement at West Wickham was founded in this period. In the high medieval period (11th–14th century) an absence of pottery from the area immediately adjacent to the present High Street south of the church (Figure 5C between "b" and "e") appears not to have been used for habitation, suggesting that there may have previously been an open green in this area, with houses sited along the northern edge of this. Both Streetly End (Figure 5(Cc)) and Burton End (Figure 5(Cd)) appear to come into existence at this date, and therefore appear later in date than West Wickham. These settlements appear small, but as few test pits were excavated, especially in Burton End, the findings to date might underestimate the level and extent of activity in this area. Overall, settlement in the parish of West Wickham in the high medieval period seems likely to have been somewhat dispersed in form, with a small village green surrounded by at least two small ends and probably several other isolated moated farms or homesteads (which were not excavated in 2013). This process of high medieval settlement expansion was abruptly arrested in the later medieval period (Figure 5D), which saw the settlement pattern particularly severely scaled back, with many outlying sites producing no pottery of later medieval date (mid-14th–mid-16th century) at all and only two pits producing more than a single sherd. Recovery at Burton End and Streetly End does not appear to have been established until after the end of the medieval period: all but three of the pits produced pottery of 16th–18th century date (Figure 5E). The pottery distribution suggests that when this robust recovery did take place, a nucleated row village developed at West Wickham near the church, where large amounts of pottery found close to the road suggests the green had been built over. At Burton End and Streetly End, the dispersed character of the settlement pattern established in the high medieval period appears to have been re-established.

3. Discussion: Deep Mapping and Wider Impacts

The four projects described above involved more than 1200 members of rural communities of place in a highly-engaged manner which involved significant contributions of time and effort by volunteers and which, as outlined above, succeeded in producing new spatial understanding of the long-term development of these four historic communities. The maps created using the data excavated from the archaeological test pits are, like all maps, partial in both senses of the word. They are selective in the types of evidence they encompass, and there are gaps in the datasets the maps do include, caused here by the incomplete nature of the archaeological record and the sampling process used which inevitably only recovers a tiny fraction of the total surviving buried data, which itself is only a fraction of that which was originally left behind by past users of the explored spaces. The practical limitations of the test pit approach when used within inhabited communities, where site selection is contingent on access which may be restricted by physical or social factors, means that some of the more fragmentary or isolated data, such as single sherds of prehistoric pottery in test pits sited far from any others, is not easily amenable to interpretation or explanation. However, each pit reveals unique new information about its individual location, and when the data from many pits are combined, analysed, and mapped, the patterns which emerge provide meaningful new perspectives on the changing use of space and place over time. The test pit excavation process thus contributes to the "depth" of mapping of settlement and landscapes at a range of scales.

Detailed discussion of how this is advancing knowledge and understanding of wider regional changes in settlement, landscape and demography [33] is beyond the scope of this paper. However, consideration will now be given to the social impact of test pit excavation projects within living communities, as the CCH projects were funded by HLF and AHRC with the aim of developing collaborations between university researchers and community groups which would enable group members and wider publics to benefit from exploring their chosen aspects of their chosen heritage. Accordingly, as well as the formal analysis, mapping and reporting of the archaeological finds, the impact of the projects was assessed by informal participant observation and formal written feedback taken from community group leaders at the end of the delivery phase in 2013 using paper forms and online surveys to elicit scalar metric assessments as well as free-text comments.

3.1. Enhancing Participant Knowledge and Engagement with Local Heritage

The projects had a significant positive impact on knowledge and engagement with local heritage within the host communities. All groups reported that the excavations had increased their own and volunteers' knowledge of their heritage, these included knowledge amongst residents of the past of their own property (gained through taking part in the excavations) and of the new wider historical maps of the village (through attending plenary social events and following digital media outputs). Overall, the extent to which involvement increased a sense of connection with heritage was rated by groups (using a scale of 1 (lowest) to 10 (highest)) at an average of 9.33. Participant observation and free-text comments alike suggested that the impact was enhanced for excavation volunteers by the fact that the investigations (firstly) took place within volunteers' own gardens and (subsequently) contributed to a wider body of new knowledge and understanding at local, regional and national level: this enhanced the sense that the outcomes at all levels were of direct relevance to volunteers. On almost every visit to each test pit, supervising archaeologists witnessed new discoveries being greeted with pleasure, interest, surprise and pride by participating community volunteers who were learning the extent, identity and significance of relics from the past recovered from beneath their lawns, playgrounds, verges and vegetable patches. Such sentiments were evident in individuals of all ages and backgrounds, and were enhanced by the intimate nature of much of the recovered evidence which strengthened the sense of connection between different people occupying the same spaces at different times. Fragments of personal items which would have been in daily use, such as tobacco pipe stems, cooking pot sherds, dolls' limbs or flint tools, which had been made or chosen, owned or used and lost or discarded by former human users of very familiar spaces, acquired a value to their finders which considerably exceeded their conventional "research value", let alone any realisable monetary value. "*It was a lot of work, but hugely rewarding and we have been overwhelmed by the interest that our project has generated*" (MLHG).

3.2. Confidence Building within Community Groups

Another significant impact of the projects was in building confidence within community groups for running community heritage projects, which will build capacity for the future. This was evident as early as the pre-submission development stage in 2012. During this period, three of the four groups expressed a desire at one point or another to withdraw from the process of devising project plans and preparing funding bids, with the main factor cited being anxiety over the complexities of the HLF application process and the possible adverse consequences of both failure and success in participating in the bid process. In each case, university support in the form of reassurance, advice, advocacy and/or problem-solving restored confidence amongst group leaders and led to all four bids not only being submitted, but in being successful in securing funding. 18 months later, when the projects had all been completed, the experience overall of running the project was rated very highly by group leaders, averaging 8875 out of 10. "*Would never have had the knowledge or confidence to embark on a project like this without the University's help and the involvement of the University was intrinsic to the success of our project... With the knowledge and skills that we have gained we now have the confidence to run similar projects in the future*" (MLHG).

3.3. Extending Social Networks within Communities

The projects also developed social networks within and between groups and communities. Group working is an integral part of an activity such as test pit digging which usually requires at least two people to complete (and can easily accommodate a dozen or more friends, family, children and neighbours). Fostering opportunities for wider social exchanges before, during and after the excavations and planning for these to be lively social gatherings was actively encouraged by the CCH team, as prior experience had shown this to be a popular and hence effective way of encouraging people to attend briefings—not necessarily a top priority for people who might be tired, busy, or simply culturally disinclined to attend "talks". Noting the addition of tea and cake to such events might seem trivial, but this encouraged volunteers to linger, relax and enjoy informally exchanging stories, information and ideas about their individual excavations, maximising the impact of the project as a nested activity which could be engaged with on a number of different levels and could lead on to other activities or interests in the future. Providing refreshments also provided opportunities for residents unwilling or unable to take part in the excavations to get involved in the project in different ways and drew in other community groups such as local Women's Institutes. All CCH community group leaders felt their test pit projects had increased their knowledge of and connectivity with people locally who were interested in their heritage, and most felt this had increased their capacity to explore other projects in the future: "*As a direct result of the project, new members have joined our Group and are now going on to research other aspects of the village's history*" (MLHG).

3.4. Instilling Skills

Another observable wider outcome of the CCH test pit projects was in the wide range of *skills* which were developed, including specific heritage-related skills as well as more broadly transferrable ones. Heritage-related skills included a range of skills in archaeological excavation and recording, creating photographic records, using archives and collections, creating archives, writing for publication and conducting local historical research. Broader transferrable skills included organising and running events, making films/audio recordings, developing webpages, using social media and working with press/media. A major driver of much of this skills development derived from a condition of HLF funding that groups should digitally disseminate the outcomes of their projects. This pushed some group members well beyond their comfort zones, but achieved very positive results in instilling skills as well as bringing the outcomes to wider audiences.

3.5. The Example of Meldreth

The Meldreth CCH project was selected as a case study by project evaluation consultants ICF GHK, commissioned by the Heritage Lottery Fund (HLF) to assess the effectiveness of All Our Stories programme [48]. ICF GHK identified 10 outcomes achieved by the Meldreth project for heritage, people and communities. The outcomes for *heritage* included the significance of the finds, the fact that these were formally reported on and the quality of the exhibition and documentation of the heritage project outcomes on the group's website. Outcomes for *people* included acquisition of new skills and new heritage knowledge, which included primary and secondary school students as well as adults and younger families. Outcomes for the *community* highlighted the inclusive and engaged character of the project: "*We were really struck by the kindness and enthusiasm of everyone. You managed the 'inclusive' event with flying colours*"..."*I have been incredibly impressed with the way the project has engaged so many and awakened so much curiosity about the village's past*"..."*it pulled the whole village together and much was learnt about Meldreth in the process*". A district and county councillor noted the "community building" achieved by the test pit project.

The enduring legacy of the project in Meldreth was evident in subsequent developments. Encouraged by the success of their test pit project, Meldreth History Group quickly developed a range of ideas for new future activity, included "...*Geophysical survey, fieldwalking and digging more test pits. We may also be interested in archaeological investigations on a larger scale if geophysics suggests that this is warranted. Other projects which may benefit from the involvement of a university student or researcher include research into the village's manorial history and the use of dendrochronology/radio carbon dating to date old timbers in buildings*". These ambitious intentions have not remained mere wishful musings [47]: In 2014 members excavated more test pits involving residents of a retirement home and a special school, as well as carrying out a geophysical survey on a manorial site excavated during the CCH project, while in 2015 they are actively planning further funding bids for larger-scale community excavations on this site. In addition, members of the group are now providing another group in the nearby village of Stapleford with equipment, advice and on-site help in running their new test pit excavation project. The CCH project thus achieved specific and measurable culture change: Meldreth Local History Group, which was previously small and had no knowledge or experience in archaeological investigation, is now growing in size and has excavation firmly established as a popular core activity which is expanding the reach and impact of the group within and beyond their community to widespread benefit and which appears securely embedded for the future.

4. Beyond CCH

CCH was a large programme which involved more than 5000 people in 28 different projects ranging from excavation to oral history, all supported simultaneously by the same small university team on a tight budget over a single 12-month period: take-up of the programme was much larger than anticipated by the funders, and resources were not available to collect written feedback from individual participants. However, similar community test pit excavation projects elsewhere, carried out using identical approaches and supported by the same university team as the four CCH projects, show that the positive *social impacts* achieved in the CCH projects are widely achieved, and here feedback provides more detail about the nature of this impact. Between 2008 and 2014 written feedback data was collected from the excavators of more than 400 test pits in 12 different communities in eastern England [33]. With each respondent typically representing between four and fifteen volunteers, the results represent the impact on perhaps 2000–5000 individuals.

4.1. Impact on Local Knowledge, Engagement and Well-Being amongst Project Volunteers

Feedback confirms the observation from CCH that test pit excavation projects have strong and diverse positive impacts on volunteers. 77% of the volunteers felt that taking part in the test pit excavations had improved their knowledge of the archaeology and history of the place they lived in, and the fact that 75% had no prior experience of archaeological excavation shows that the activity reaches well beyond those who are already engaged. The impact of participation on levels of engagement with heritage was also explored in feedback, which showed that 84% felt the experience had led them to feel *more* engaged with their local heritage. Nearly the same number (83%) said that as a result they would take more interest in their local heritage *in the future*. Also looking to the future, 96% of respondents said they would recommend the experience to others. A follow-up question on the feedback forms invited those replying "yes" to this "recommend?" question to add free-text comments explaining *why* they would recommend the activity to others. Responses (taken here from a project in Nayland in Suffolk in 2012 [52] which typifies responses from other projects), illuminates the nature of the impact of this activity on participants as a number of recurring themes become apparent.

The first of these is the sheer enjoyment people had in the activity: "*It's such good fun*" (PR); "*I noticed too a lot of young, inexperienced people taking part who were clearly hugely enjoying themselves*" (KH). This enjoyment is reflected in aggregate data: 97% of respondents rated the community test pit experience as "good" or "excellent". The activity is also appreciated for its educational value: "*It was a great educational experience, and enlightening to find out the history of Nayland. I developed good practical skills*" (KP); "*Good for all ages to learn about the history of an area (particularly their own)*" (DD); "*A great learning experience*" (RH); "[I enjoyed] *finding out about the history of the local area and about history/archaeology generally*" (PR). The social aspect of the excavations is repeatedly specifically listed as a reason people would recommend the activity: "*Brings strangers together working in teams*" (AC); "*I loved working with different people*" (PR); "*Getting to meet new and interesting people especially within the village*" (ME); "*...A good way to...meet new people when working as part of a team*" (GE). This positive social impact is often described in ways which show its potential to generate enduring bonds between people and within the host communities: "*Being part of a team creates and promotes social support and knowledge of each other*" (LF); "*Promotes a common interest and helps the village to bond*" (LF); "*It is very social, you begin to feel one of a team, one of a community*" (DC). Importantly in this respect, the excavations also attract people from a range of backgrounds: "*Good for bringing together people of all ages*" (RH). This hints at the range of ways in which participation contributes to personal well-being, but building social links has the capacity also to strengthen and foster resilience within communities.

The opportunity the excavations provide for close contact with past is also frequently commented on: "*It was a great way to spend the weekend outdoors, learning about history in a hands-on and up close and personal manner*" (KP); "*It is really good to touch and find things that no one will have seen for many years*" (SW); "*It is very satisfying to get some hands on experience*" (SW); "*Being actively involved is very different to just reading about it*" (DC). A feeling that taking part changed volunteers' perspectives on their community and its connection with its past was also often cited: "*It's a really mind-opening experience*" (PR); "*Helps to give you a sense of the context of your community*" (RH); "*Archaeology is about our history, it belongs to all of us*" *(DC)*; "*Encourages you to think about the people who lived here before us and how they lived*" (SM). The knowledge-enhancement aspect of the excavations were also appreciated as extending beyond individuals, being satisfying because they generating new knowledge for others: "*It helps to improve knowledge of Nayland's history and it was a very interesting experience as it encourages a more active interest in history and archaeology*" (GE); "*Fascinating insight into local history. Contributing to a national grid of archaeological information*" (MH). Creating stronger perceptual links between people and their social and historical landscapes has evident synergies with deep mapping.

While the "discovery" and "social" impacts of participation can be seen as distinct achievements, they are appreciated by volunteers as entirely integrated, typified by comments such as: "*I loved working with different people and helping to contribute to adding to the knowledge of this region*" (PR); "*It was great to be part of something that involved so many of the village and that feeds into a much bigger picture*" (RH).

4.2. Impact of Participation on Volunteer Skills

A framework has been developed [53] for assessing the ways in which taking part in test pit excavations instils transferrable skills in teenagers participating in the aspiration-raising Higher Education Field Academy (HEFA) programme [54] where they carry out exactly the same archaeological procedures as the CCH project. A review of the assessment data [55] showed that in 2012–2014, 77% of nearly 1000 HEFA students felt participation had developed their verbal communication skills (with assessors grading 74% of students at A*–B in this skill); 87% felt it had developed structured working skills (76% graded A*–B by assessors); 79% developed creative thinking skills (66% graded A*–B); 78% developed reflective learning skills (72% A*–B); 86% developed perseverance skills (88% A*–B) and 87% developed team working skills (82% A*–B).

Participation in test pit excavation projects has also been shown to help people who face particular challenges. In 2012, collaboration with Cambridgeshire charity Red2Green [56] involved adults affected by autism in tailored test pit projects intended to increase participants' interest in archaeology and local history and to develop their teamwork and communication skills [57]. In 2012, 15 volunteers who belonged to Red2Green's "Aspirations" life-skills building group took part in test pit excavations in Swaffham Bulbeck for up to three days. Most excavated alongside pupils attending the local primary and secondary schools, and subsequently worked with an artist to create display items drawing on their experience including patchwork quilts and decorated ceramic tiles. Feedback on the excavation element of the project ([57], Appendix) showed that 73% of the Red2Green participants rated the experience very highly, with 85% saying they would like to get involved with practical archaeology projects again in the future. 69% said they would like more chances to discuss archaeology or local history with others, supporting the observation by Red2Green staff that the learners had demonstrated a *"noticeable increase of confidence when talking to new people"*. One parent commented enthusiastically that her son, who did not usually talk much about his day "*would not stop talking about the project!*" highlighting not only his enjoyment, but also the impact the project had on his ability and desire to vocalise his experiences. These are notable outcomes given the difficulties these Red2Green clients normally encounter when communicating verbally with others.

4.3. Enduring Legacies

The long-term outcomes of sustained involvement with community test pit excavation projects can be seen at Pirton in north Hertfordshire, ([33], pp. 325–26) (which lies immediately adjacent to the CCH 2013 village of Shillington). Test pit excavation at Pirton started in 2007 as a school HEFA project with a small group of teenagers completing just five pits ([58], pp. 52–53). It might have stopped there, but when local residents heard of the results from other HEFA communities where more than 30 pits had by then been excavated, they were inspired to take up the challenge. Over the course of a series of biannual weekend community excavations, scores of local residents completed more pits and cumulatively gained the skills and knowledge to be able to work effectively without professional archaeological supervision, although contact with the university was maintained throughout. Excavations continued for eight years, and by 2015 more than 100 pits had been completed in Pirton, more than has yet been achieved in any currently occupied rural settlement. In 2015, aware of the academic context and recognising the wider importance of their results, the Pirton community decided to move onto the next stage, as announced in coordinator Gil Burleigh's Facebook post on 17 May 2015: "*Sad and exhilarating weekend! Dug 115th test pit in Pirton, found western extent of 13th century manorial building first identified in 2013, plus 15th–18th century barn, at Bury End. That's the end of the project though—time to write it all up, everything since 2007. Thanks to everyone who has volunteered and contributed over past eight years...*" [59]. There is clearly ample potential here for creating a deep map of Pirton which encompasses the intangible contributions of time, energy, local knowledge and ideas made by scores of local volunteers as well as the tangible artefactual evidence spanning thousands of years.

Indeed, the potential of this stage of community test pit excavation projects can be seen in the results from Great Bowden, whose local historical society [60] collects village photographs and other memorabilia and publishes local historical studies [61]. In 2013–2014 the society secured funding for a community test pit excavation project and completed 30 pits across this Leicestershire village. The results were analysed by the same University of Cambridge team as the CCH projects in collaboration with the Great Bowden Heritage and Archaeology Group, whose members contributed accounts of the local and oral history of each of the excavated sites which were included in the formal excavation report [62]. As a result, site-by-site analyses and time-deep maps consequently enmesh local history, archaeology and personal memory.

5. Conclusions

A deep map can be defined as "a finely detailed, multimedia depiction of a place and the people, animals, and objects that exist within it and are thus inseparable from the contours and rhythms of everyday life...simultaneously a platform, a process and a product...a way to engage evidence within its spatiotemporal context and trace pathways of discovery...framed as a conversation and not a statement" ([13], pp. 3–4)—an answer to the challenge of holistically comprehending space, time, people and place. Archaeology is often overlooked in the scholarly literature on deep mapping, but has the capacity to contribute significantly to the creation of deep maps, not only by adding greater time-depth but also, though the palpable, grounded, on-site physicality of its evidence base, strengthening the connections between people, space, time and map.

The data presented here, from CCH projects and other similar programmes, show that archaeological test pit excavation within currently occupied rural settlements—today's villages, hamlets, farms and small towns—generates deep maps which elicit unique new spatiotemporal perspectives. They can show how different areas within today's inhabited places were used at different times in the past, revealing the often turbulent ebb and flow of human endeavour concealed underneath the for-now quiescent lawns of 21st century rural villages, showing observers which spots have been continuously inhabited for a millennium or more, which are new expansions within the last couple of centuries, and which have for generations oscillated between habitation, abandonment and re-occupation. The data also glimpse more personal histories—the ravaging of Shillington by the demographic catastrophes of the 14th century; the aspirational consumption of the manorial lords of Meldreth, evident in the exotic mirror found at Flambards manor; and the simpler everyday pleasures of rural life evidenced by numerous finds of fragments of clay tobacco pipes and children's toys. These data create multi-media maps whose depth is both temporal and psychological, derived from the living landscape, imbued with the physicality of the tangible nature of the recovered evidence and dynamically capable of growing and changing as new data is added or ideas change.

When the excavations are carried out by residents of those communities, they acquire depth in another dimension as people search, for themselves, within their own familiar places, for evidence from the seemingly remote past. The process of data recovery involves participants in performance of the processes of excavation which are both structured and improvised: each team will have carried out the same actions, but in their own way in response to the unique qualities of their site, their team. Enacting this performance creates bonds between participants, whether they have excavated on the same site or different ones. Furthermore, the palpable physicality of the recovered archaeological evidence creates an umbilical link between present inhabitants and their predecessors that can be appreciated both intellectually and empathetically, and which can engage, meaningfully, people with very different levels of prior knowledge and interest[1].

Community test pit excavation responds to the notion of deep mapping, perhaps even epitomises it, as it involves people in exploring evidence which is neither finite nor constrained, simply recovering whatever is present beneath a single square metre of ground and weaving the resulting material into a narrative about a place. It combines subjective, experiential and performative inputs with objective, standardised frameworks of method and analysis to create multi-media outputs which are academic, personal and capable of dynamic evolution in the future.

Reference to the future raises another important point. Given the diverse academic and social benefits of this approach to deep mapping, it is clear that extending this activity is desirable. To do this, it is crucial that the knowledge and understanding which these deep maps represent should not be restricted, within either academia or local communities, not least because dissemination will help fulfil potential. Dissemination is not, however, automatically consequential on the primary activity and needs to be specifically planned for. All the CCH projects presented here returned the outcomes to host communities in in public events where finds could be examined, new maps presented and the results discussed (dissemination was an explicit requirement of the HLF, who prioritise the social impact of funded projects). Some groups created permanent physical displays, and all added content to websites. These disseminations, especially the events, were popular and highly impactful locally,

---

1 This is illustrated by an e-mail received from one participant in On Landguard Point "I was given the silver rifle charm yesterday to bury in a test pit my team dug in my back garden. As I lay the rifle in the bottom of my pit I had a sudden rush of emotion: my son is in Afghanistan at the moment serving in our armed forces. I snatched the rifle from the pit and put it back in the envelope I had freed it from two minutes earlier. My promise to you is, in October when my son returns safely home I will place this charm in the ground at a depth of 500 mm. My son, and my family, make my Home…" (PP, excavated on Dig and Sow in Potton, June 2012) [64].

but nonetheless, their impact will dissipate over time. Public broadcast disseminated the outputs of community test projects similar to CCH in Kibworth (Leicestershire) in *Story of England* in 2010 [63]; and in six different communities in eastern England in *On Landguard Point* (an Arts Council-funded commissioned for the London 2012 Olympic Games' Cultural Olympiad [64]). Such dissemination massively increased the reach of these projects: most of the CCH community partners had been inspired by *Story of England.* Realistically, however, such broadcast opportunities are rarely available, and if deep mapping is to be advanced and embedded within wider publics in a truly sustainable way, a framework for continued exploration is needed which stimulates *mass engagement* at a *local level.* This seems ambitious, perhaps, but experience of community test pit excavation projects suggests that it is achievable, and routes to such achievement, through publicly engaged research involving community groups and school curricula, are currently being explored by this author.

Viewed in the context of deep mapping, it is salient that community test pit excavation projects create performative and perceptual links between present and past communities, as people are connected with both the process and product of following horizontal and vertical pathways of discovery about their own locales, while simultaneously generating new evidence which can be mapped within its spatio-temporal context and aggregated to advance wider understanding of changing places in time. That such projects simultaneously nurture personal well-being and build social capital for the future within those same communities demonstrates the wider value of such participatory community-engaged research.

Acknowledgments

The CCH project was funded by the Arts and Humanities Research Council and the Heritage Lottery Fund, and their support is gratefully acknowledged. The Principal Investigator and Archaeological Director for CCH (and for HEFA, *Story of England* and *On Landguard Point*) was Carenza Lewis. Alexander Pryor, Britt Baillie, Catherine Ranson, Matthew Collins and Sue Oosthuizen provided archaeological advice and supervision to excavating volunteers and coordinators within the community groups and Clem Cooper provided essential administrative support to the PI. Paul Blinkhorn, Lawrence Billington and Vida Rajcovaca respectively reported on the pottery, worked flint and animal bone from the excavations. Volunteers locally who took part in the excavations are far too numerous to name here individually, but special acknowledgement is due to Kath Betts and Joan Gane (Meldreth Local History Group): Janet Morris and Andrew Morris (West Wickham Local History Group); Derek Turner (Shillington Local History Society); Mike McCarthy, Ann Mitchell and Colen Lumley (Toft Historical Society), who suggested and ran the projects within their local communities, without whom none of the CCH test pit projects presented here would have happened. Gil Burleigh, Helen Hofton, Andora Carver and Rosemary Culkin were responsible for coordinating test pit excavation projects at Pirton, Nayland and Great Bowden, also discussed in this paper.

Conflicts of Interest

The author declares no conflict of interest.

References and Notes

1. Heat-Moon, William Least. *PrairyErth: A Deep Map*. Boston: Houghton Mifflin Company, 1991.
2. Wydeven, Joseph J. "Review of PrairyErth: A Deep Map." *Great Plains Quarterly* 1 (1993): 133–34.
3. Pearson, Mike. *In Comes I*. Exeter: University of Exeter Press, 2006.
4. Kerouac, Jack. *On the Road*. New York: Viking, 1957.
5. Bryson, Bill. *A Walk in the Woods: Rediscovering America on the Appalachian Trail*. London: Random House, 1997.
6. Wood, Michael. *The Story of England*. London: Penguin Books, 2010.
7. McMaster, Ida, and Kathleen Evans. *Mount Bures: Its Lands and Its People*. Bures: Privately printed, 1996.
8. Cooper, Ashley. *The Long Furrow: 2000 Years of Rural History along the Suffolk-Essex Border*. Bulmer: Bulmer Local History Group, 2009.
9. Hallman, Robert. "Thundersley and Daws Heath: A History." 2015. Available online: http://www.hadleighhistory.org.uk/page/thundersley_and_daws_heath_a_history (accessed on 5 July 2015).
10. Hoskins, William George. *The Making of the English Landscape*. Leicester: Penguin Books, 1955.
11. Trubshaw, Bob. "Book Review—Psychogeography." *Time and Mind: The Journal of Archaeology, Consciousness and Culture* 2 (2009): 87–92.
12. Ashmore, Wendy, and A. Bernard Knapp, eds. *Archaeologies of Landscape: Contemporary Perspectives*. Oxford: Blackwell, 1999.
13. Bodenhamer, David J., John Corrigan, and Trevor M. Harris. *Deep Maps and Spatial Narratives*. Bloomington: Indiana University Press, 2015.
14. Knowles, Anne Kelly. *Past Time, Past Place: GIS for Historians*. New York: ESRI Press, 2002.
15. Wheatley David, and Mark Gillings. *Spatial Technology and Archaeology*. London: Taylor and Francis, 2002.
16. Knowles, Anne Kelly, and Amy Hillier. *Placing History: How Maps, Spatial Data, and GIS Are Changing Historical Scholarship*. New York: ESRI Press, 2008.
17. Schiavini, Cinzia. "Writing the Land: Horizontality, Verticality and Deep Travel in William Least Heat-Moon's PrairyErth." *Rivista di Studi Americani* 15–16 (2004–2005): 93–113.
18. David, Bruno, and Julian Thomas. *Handbook of Landscape Archaeology*. Walnut Creek: Left Coast Press, 2008.
19. Petts, David. "Archaeology and Psychogeography." Available online: http://outlandish-knight.blogspot.co.uk/2011/12/archaeology-and-psychogeography.html (accessed on 7 September 2015).
20. Ethington, Philip, and Nobuko Toyosawa. "Inscribing the Past: Depth as Narrative in Historical Spacetime." In *Deep Maps and Spatial Narratives*. Edited by David J. Bodenhamer, John Corrigan and Trevor M. Harris. Bloomington: Indiana University Press, 2015, pp. 72–101.

21. Rohl, Darrell J. "Chorography: History, Theory and Potential for Archaeological Research." In *TRAC 2011: Proceedings of the Twenty-First Theoretical Roman Archaeology Conference*. Oxford: Oxbow Books 2012, pp. 19–32.
22. Aldenderfer, Mark, and Herbert D. G. Maschner, eds. *Anthropology, Space and Geographic Information Systems.* Oxford: Oxford University Press, 1996.
23. Conolly, James, and Mark Lake. *Geographical Information Systems in Archaeology.* Cambridge: Cambridge University Press, 2006.
24. Byrne, Denis. "Counter-mapping in the Archaeological Landscape." In *Handbook of Landscape Archaeology*. Edited by Brund David and Julian Thomas. Walnut Creek: Left Coast Press, 2008, pp. 609–16.
25. Bradley, John J. "When a Stone Tool is a Dingo: Country and Relatedness in Australian Aboriginal Notions of Landscape." In *Handbook of Landscape Archaeology*. Edited by Brund David and Julian Thomas. Walnut Creek: Left Coast Press, 2008, pp. 633–37.
26. Tilley, Christopher Y. *A Phenomenology of Landscape.* Oxford: Berg, 1994.
27. Know Your Bristol. "Exploring local heritage and culture through the eyes of Bristol communities." 2015. Available online: http://knowyourbristol.org/about-us/ (accessed on 17 July 2015).
28. Skeates, Robin, Carol McDavid, and John Carman, eds. *The Oxford Handbook of Public Archaeology*. Oxford: Oxford University Press, 2012.
29. Cambridge Community Heritage. Available online: http://www.access.arch.cam.ac.uk/communities/cch (accessed on 30 June 2015).
30. Arts and Humanities Research Council Connected Communities. Available online: http://www.ahrc.ac.uk/research/fundedthemesandprogrammes/crosscouncilprogrammes/connectedcommunities/ (accessed on 7 September 2015).
31. Heritage Lottery Fund. "All Our Stories: Heritage Lottery Fund launches new funding programme." 2012. Available online: https://www.hlf.org.uk/about-us/media-centre/press-releases/all-our-stories-heritage-lottery-fund-launches-new-funding (accessed on 5 July 2015).
32. Lewis, Carenza. *Cambridge Community Heritage*. Cambridge: McDonald Institute for Archaeological Research, 2013, submitted for publication.
33. Lewis, Carenza. "The Power of Pits: Archaeology, outreach and research in living landscapes." In *Living in the Landscape*. Edited by Katherine Boyle, Ryan J. Rabett and Chris O. Hunt. Cambridge: McDonald Institute for Archaeological Research, 2014, pp. 321–38.
34. Gerrard, Christopher M., and Michael Antony Aston. *The Shapwick Project, Somerset: A Rural Landscape Explored.* Leeds: Society for Medieval Archaeology, 2007.
35. Jones, Richard, and Mark Page. *Medieval Villages, Beginning and Ends*. Macclesfield: Windgather Press, 2007.
36. Beresford, Maurice Warwick, and John Hurst. *The Lost Villages of England.* London: Alan Sutton, 1954.
37. Hooke, Della, ed. *Deserted Medieval Villages: A Review of Current Work.* Oxford: Oxford University Committee for Archaeology, 1985.
38. Dyer, Christopher, and Richard Jones, eds. *Deserted Villages Revisited.* Hatfield: University of Hertfordshire Press, 2010.

39. Christie, Neil, and Paul Stamper, eds. *Medieval Rural Settlement. Britain and Ireland AD 800–1600.* Oxford: Windgather and Oxbow Books, 2012.
40. Lewis, Carenza, and Britt Baillie. "Archaeological Test Pit Excavations at West Wickham, Cambridgeshire." 2013. Available online: http://www.access.arch.cam.ac.uk/reports/cambridgeshire/west-wickham/2013 (accessed on 7 September 2015).
41. Lewis, Carenza, and Alex Pryor. "Archaeological Test Pit Excavations at Meldreth, Cambridgeshire." 2013. Available online: http://www.access.arch.cam.ac.uk/reports/cambridgeshire/meldreth/2013 (accessed on 7 September 2015).
42. Lewis, Carenza, and Alex Pryor. "Archaeological Test Pit Excavations at Shillington, Bedfordshire." 2013. Available online: http://www.access.arch.cam.ac.uk/reports/bedfordshire/shillington/2013 (accessed on 7 September 2015).
43. Lewis, Carenza, and Alex Pryor. "Archaeological Test Pit Excavations at Toft, Cambridgeshire." 2013. Available online: http://www.access.arch.cam.ac.uk/reports/cambridgeshire/toft/2013 2013 (accessed on 7 September 2015).
44. Lewis, Carenza. "Test pit excavation within occupied settlements in East Anglia in 2013." *Medieval Settlement Research* 29 (2013): 78–89.
45. Lewis, Carenza. "Disaster Recovery—New archaeological evidence for the long-term impact of the 'calamitous' 14th century." *Antiquity*, 2015, forthcoming.
46. Betts, Kathryn. "Meldreth Local History Group." 2007. Available online: http://www.meldrethhistory.org.uk/page_id__23.aspx (accessed on 5 July 2015).
47. Meldreth History. "2013 Test Pit Project." 2013. Available online: http://www.meldrethhistory.org.uk/category_id__103.aspx (accessed on 5 July 2015).
48. Heritage Lottery Fund. "All Our Stories Evaluation." 2015. Available online: http://www.hlf.org.uk/all-our-stories-evaluation (accessed 5 July 2015).
49. Shillington History Society. Available online: http://www.shillington-history.org.uk/ (accessed on 5 July 2015).
50. Toft Historical Society. Available online: http://www.tofthistoricalsociety.org.uk/aos/ (accessed on 5 July 2015).
51. Linking with Your Local Community. "West Wickham and District Local History Club." Available online: http://www.cambridgeshire.net/organisation/west-wickham--district-local-history-club/21531.aspx#.Va1-F0ZGRQJ (accessed on 5 July 2015).
52. Lewis, Carenza, Catherine Ranson, and Alex Pryor. "Archaeological test pit excavations at Nayland, Suffolk." Available online: http://www.access.arch.cam.ac.uk/communities/past/managing-a-masterpiece/nayland/Nayland_2012_Report_Final.pdf (accessed on 18 August 2015).
53. Johnson, Martin, and Carenza Lewis. "'Can you dig it?' Developing an approach to validly assessing diverse skills in an archaeological context." *Journal of Vocational Education & Training* 65 (2013): 177–92.
54. Access Cambridge Archaeology. "Higher Education Field Academy." Available online: http://www.access.arch.cam.ac.uk/schools/hefa (accessed on 5 July 2015).
55. Lewis, Carenza. "'Cooler than a trip to Alton Towers': Assessing the impact of the Higher Education Field Academy 2005–2013." *Public Archaeology* 13 (2014): 295–322.
56. Red2 Green. Available online: http://www.red2green.org/ (accessed on 5 July 2015).

57. Lewis, Carenza, Catherine Ranson, and Alex Pryor. "Archaeological Test Pit Excavations at Swaffham Bulbeck, Cambridgeshire." Available online: http://www.access. arch.cam.ac.uk/reports/cambridgeshire/swaffhambulbeck/2012 (accessed on 5 July 2015).
58. Lewis, Carenza. "Test pit excavation within occupied settlements in East Anglia in 2007." *Medieval Settlement Research Group Annual Report* 22 (2007): 48–56.
59. Burleigh, Gilbert. "Facebook post". Available online: https://www.facebook.com/gil.burleigh?fref=ts (accessed on 5 July 2015).
60. Great Bowden Heritage and Archaeology. Available online: http://www.greatbowdenheritage.btck.co.uk/ (accessed on 5 July 2015).
61. Great Bowden Historical Society. Available online: http://www.greatbowdenhistoricalsociety.btck.co.uk/Publications (accessed on 5 July 2015).
62. Lewis, Carenza, Catherine Ranson, and GBHAG. "Archaeological Test Pit Excavations in Great Bowden, Leicestershire, 2013 & 2014." 2014. Available online: http://www.access.arch.cam.ac.uk/reports/Leicestershire/greatbowden/2014 (accessed on 5 July 2015).
63. BBC2. "Story of England (TV series)." 2010. Available online: http://www.imdb.com/title/tt1735365/ (accessed on 18 August 2015).
64. Lewis, Carenza. "On Landguard Point." In *Archaeology at Cambridge 2011–2012*. Cambridge: McDonald Institute for Archaeological Research, 2012, p. 54.

Long Street: A Map of Post-Apartheid Cape Town

Giovanni Spissu

Abstract: No map fully coincides with the territory it represents. If the map and territory do not coincide, what can the map capture of the territory? According to Bateson, the answer is its differences. Drawing from Gregory Bateson's ideas, we can envision an ethnographic representation of the city through which we can represent the urban territory through the different ways its inhabitants perceive it. In this article, I describe the process that led me to build a map of post-apartheid Cape Town from Long Street. I took inspiration from Bateson's book *Naven* and compared it with the District Six Museum map in Cape Town with the objective of representing post-apartheid Cape Town through its differences.

Reprinted from *Humanities*. Cite as: Spissu, G. Long Street: A Map of Post-Apartheid Cape Town. *Humanities* **2015**, *4*, 436–456.

1. Summary

"Differences are the things that get onto a map" ([1], p. 465).

In 1931, the Polish-American philosopher Alfred Korzybsky used the dictum "*The map is not the territory*" ([2], p. 8) to sum up the principle that no type of representation could be considered a true replica of the represented object. Korzybsky stated that human knowledge of the world was limited by our neuronal system and the type of language we have developed. For this reason, according to the philosopher, access to reality is always filtered and mediated by different forms of abstraction (which are often taken for reality). According to this principle, it follows that no map, or other type of representation, whether artistic or scientific, could ever correspond to the object to which it refers. Korzybsky's formula is particularly applicable to the world of ethnography and social sciences in general, in which no type of representation can capture the object of study in its entirety or totality. Any representation, and particularly ethnographic representation, will always be partial and limited.

But if no map ever corresponds to the territory to which it refers, what separates them? What can the map capture of the territory? According to Bateson, what separates the map from the territory are their differences. He wrote:

> "We know the territory does not get onto the map. That is the central point about which we here are all agreed. Now, if the territory were uniform, nothing would get onto the map except its boundaries, which are the points at which it ceases to be uniform against some larger matrix. What gets onto the map, in fact, is difference, be it a difference in altitude, a difference in vegetation, a difference in population structure, a difference in surface, and so forth. Differences are the things that get onto a map" ([1], p. 465).

But if there are differences that separate and (join) map and territory, what are the differences? Bateson considered "difference" to be "a very peculiar and obscure concept." ([1], p. 465). He said that difference was not to be sought only in physical characteristics (color, texture, shape, *etc.*) ([1], p. 467) of the representing subject and represented subject. The origin of differences cannot be sought in the map or in the territory but in the relationship between the two. Map and territory cannot be considered separate, independent realities; they are connected entities in a reciprocal relationship expressed through differences.

Many scholars have recently thrown into doubt the conception of the map as a static, unified representation and have started to recognize it as a dynamic, process-based entity. The discussion has moved beyond questioning the map as a faithful replica of reality or presumed bearer of scientific objectivity; it now pertains to the type of relationship it can establish with the territory. James Corner, for example, turns over the common conception of the map as a representation of the territory and maintains that maps should be seen within a single interpretative process. According to Corner, the territory does not proceed the map, but it is the space that becomes territory through the process of mapping. Corner notes that "mapping involves a process of gathering, working, reworking, assembling, relating, sifting....speculating and so on (that) allow certain sets of possibility to become actual" ([3], p. 228). According to him, maps should not be seen only as representations of the territory but as makers of it: "Maps remake territory over and over again, each time with new and diverse consequences." ([3], p. 213) We cannot therefore think of the two agents independently; map and territory are co-constructed. Whereas on the one hand we cannot conceive of the map outside of its references, on the other, the territory is activated by the practice of mapping.

Taking inspiration from these sources and drawing a parallel between Bateson's epistemology of differences, we can envision a kind of ethnographic representation that seeks to represent the territory through its differences, meaning a multiple production of meanings arising from the dialectical relationship between map and territory. In this article, I describe the process that led me to build "Long Street: A Map of Post-Apartheid Cape Town." Drawing from Bateson's ideas, I sought to build a map of Cape Town that could represent the city through its differences. In my work, I borrowed the different viewpoints of the city's inhabitants and their way of attributing meaning to the urban spaces, placing them in relation to their personal experiences. The map in this sense is intended not as a mere representation of the city, but a multiple production of maps through which the people of Cape Town come into the relationship with the urban territory in different ways.

In March 2011, I went to Cape Town with the goal of investigating the processes of signifying urban spaces in post-apartheid Cape Town. In my work, I chose to focus on Long Street, one of Cape Town's oldest streets in the heart of its central business district. One of my primary objectives was to create a map of the street through which to represent the relationship between the street and its regulars. Exploring Long Street, its buildings, bars and alleys, and, most importantly, coming into contact with its locals, I realized how the street could not be taken as a homogeneous entity, a cohesive single "whole," uniform in its parts. On the contrary, it showed itself as a multiple, heterogeneous entity, made up of different ways to be experienced, understood and perceived. John Allen noted that "we cannot grasp the city as a whole, precisely because it is not a singular entity" ([4], p. 53), but that, on the contrary, it should be seen as a collection of "many worlds." Likewise, Long Street can be considered not just a world unto itself, but rather a constellation of different worlds made up of different languages, socio-cultural backgrounds, lifestyles and ways of perceiving the city by its regulars. The discovery of Long Street's multiplicity presented me with a difficult choice in how to represent it. At first, I saw only two solutions: I could represent it in a map in which it appeared as a single, cohesive, uniform whole within its borders; or I could concentrate on the perspective of its individual regulars. While the first solution would have let me look at the street as a whole, this strategy ran the risk of suffocating and flattening the different voices of its regulars in a single narrative. On the other hand, if I focused on the perspectives of individuals, I might fall into a reductionist trap, in which attention on individual points would prevent me from having a view of the whole.

A third solution was suggested by a happenstance visit to the District Six Museum in Cape Town and the discovery of the District Six map on its first floor. The map portrays the topographic arrangement of District Six, which was entirely demolished in the 1960s and 1970s on the orders of the National Party government; between 55,000 and 65,000 people were forcefully expelled from here. The map's most salient feature is that it was built by ex-residents who decorated it with poetry and thoughts through which they recounted their relationship with the district. The District Six map can be seen as an intersection of different ways of "mapping" a territory that came out of the different ways its residents have of perceiving it. As I looked at the District Six map, I realized that I could compare it to one of Gregory Bateson's early books, *Naven* [5]. The shared trait of the two works, so different in meaning and styles of representation, is in the multiplicity of viewpoints. In *Naven*, the young Bateson adopted a method that involved multiplying viewpoints on a single object of interest (the ritual of sexual inversion of the Iatmul people of New Guinea). He considered the overlapping of partial perspectives as a tactic of investigation; in the District Six map, the district is told through the intersection and juxtaposition of its ex-residents' perspectives. In both works, what might seem a simple narrative style underpins an epistemological criticism of every method that had synthesis as its organizing system, in order to shift the discussion to a higher level in terms of unity, homogeneity and identity. The multiplication of viewpoints and overlapping of representations in the two works seems to come out of their shared awareness of the unrepresentability of the object of interest (in its entirety). Drawing from these works, I tried to think of a kind of ethnographic representation of the city made up of a multiplicity of viewpoints and to envision a map of the city that could represent the different perspectives and voices of its inhabitants, in addition to that of the researcher. Taking inspiration from the works mentioned and from contributions from art (surrealist deambulations, situationism, the figure of the urban *flâneur*) and social scientists (Corner [3], Crapanzano [6], De Certeau 1984 [7], Ingold 2000 [8], Irving [9,10]), I sought to construct a number of urban pathways in which Long Street's people mapped the urban territory, relating it to their personal experiences in its different spaces. By overlapping and intersecting the different ways of observing, perceiving, interpreting and depicting the city, I built a representation of it.

2. Long Street: Street of Differences

In the introduction to *Imagining the City* [11], Steve Field describes different positions from which Cape Town can be observed. He gives the example of two rhetorical images through which

Cape Town is represented for tourists (that of the gateway to Africa and the multicultural Cape Town of the Rainbow Nation.) The first scene includes: "Devil's Peak, Table Mountain, Lion's Head and Signal Hill, and the city center located between the mountains and the bay" ([11], p. 6). The second includes: "The center of the view is of sprawling suburbs from the edge of Devil's Peak and the Cape Flats, reaching as far as the outer limits of Khayelitsha. The view is framed at the edges by Table Bay to the north and False Bay to the south, and is best observed from the vantage point of Rhodes Memorial, the monument erected in honour of the architect of imperial conquest, Cecil John Rhodes" ([11], p. 6).

Each of these images reveals different aspects of Cape Town, but, as Field notes, these images both reveal and conceal. One of the most interesting aspects of the first image is the "hidden perspective of the viewer," which is the point where the image is taken: Blouberg Beach. During Apartheid, this beach was a "white only area." "For the majority of Capetonians classified colored, African and Asian, it was for many decades one amongst many sites of racist exclusion by the apartheid government" ([11], p. 12).

Through this example, Field wants to suggest that no representation (written or visual) of the city can be considered complete and comprehensive. Every view of the city is necessarily selective and cannot capture it fully in its complexity. As a representation reveals something of the object it wants to show, it conceals something of it too. This example is particularly significant in studying South African cities in the post-apartheid period; as Mbembe and Nuttal have noted, they were often observed through the lens of a single development paradigm ([12], p. 8), often more concerned "with whether the city is changing along vectors of institutional governance, deracialisation of service provision and local politics than with citiness as such" ([12], p. 9).

Nonetheless, few studies have focused on the different ways of perceiving the urban territory by regular people. Recently, among many scholars of post-Apartheid phenomenon, the need has arisen to abandon the broad generalizations that framed the relationship between the public and urban places within and adopt more complex approaches that can observe processes of signifying urban places, keeping aware of their discontinuous and contradictory character. (Mbembe & Nuttal [13], Murray [12], Field, Meyer, Swanson, [11]). Field, for example, suggests a reading of Cape Town in which the city is observed through meanings produced from below, through the memories, imaginations, and thoughts of regular people. "We assert the centrality of people's creative attempts to construct, contest and maintain a material and emotionally secure sense of place and identity in Cape Town" (Field [11], p. 7). The author's choice is in keeping with a tendency shared by many South African authors, who have shown distaste for the tendency to see the post-apartheid city in a monologic and verticalistic view. Drawing from these works, in my project I intended to explore Cape Town through a multiple perspective model of exploration, observing it through the different viewpoints of its inhabitants.

In order to pursue my objectives, I decided to take Long Street as a starting point for my research. One of Cape Town's oldest streets, it is known as one of the most diverse, mixed places in the city. Since apartheid, Long Street has been considered a tolerant, liberal space, in some ways separate from the rigid social control of other areas of Cape Town. Despite the many restrictions, here people of different races continue to live in a relatively cordial atmosphere, far from the state of tension that pervaded other areas of the city. In the post-apartheid period, the street continued to be a crossroads of people of different races, geographic origins, and economic and socio-cultural backgrounds. The street is seen as divided into two parts: Lower Long Street and Upper Long Street. The lower part includes many offices and the headquarters of major companies and national and international banks. The lower part has many shops, restaurants, night clubs, coffee shops, sex shops, small hotels and bed and breakfasts.

Every evening, and especially on weekends, this part of the street is invaded by people who come from different areas of the city and from other parts of the world. There are students from Rondebosch and Observatory, arriving by taxi or private car. I met Mike, a young German man, who had lived in South Africa for several years, having done his PhD and then his post-doc there. He had married a black woman from Zimbabwe and explained to me that she preferred frequenting areas like Observatory and Long Street "where mixed couples are not the exception."

There are young people from the townships who have come to Long Street by minibus and will not return home before the next morning, when the first public transport leaves for Langa, Kayelithsa or Nyanga. Some wear pins and symbols which extol the African National Congress and give a clear indication of their ideological stance. Many of them are poor, often unemployed, and others have menial jobs; Shumbuko, a young Zulu living in Kayelitsha explained to me that being in Long Street for the evening and not in a township bar or shebeen is already a real privilege:

> "I come here often because I work here, but there are guys who, even though they live in Cape Town, have never been to Long Street."

Shumbuko's dream is to "make loads of money." His role-models are the so-called "black diamonds," prototypes of the new upwardly mobile black middle class. Shumbuko explains that they are "the idols and forbidden dream of the boys and girls from the townships and the nightmare of the new white working class." In this area of the city, it is not unusual to see them hurtling past in high-powered Mercedes and Audis, dressed in pure silk shirts and often with ostentatious jewelry.

The tourists who stay in Long Street's bed and breakfasts usually move around in groups, discussing how to spend the evening under its porticos. One of them has a T-shirt that says "Googletu," ironically combining the name of the most-visited search engine in the world with that of the city's poorest and most marginal township: Guguletu. This wordplay humorizes, defuses and distances the destitution and violence of Cape Town's urban slums.

Upper Long Street is also somewhere you can find a job; many taxi drivers have found work on Long Street. Often they come from other African countries. Lying in wait outside the busiest bars and clubs, Long Street's taxi drivers study their customers and, over time, can develop into "spontaneous anthropologists," adept at spotting potential regular customers at first sight. Desmond, a refugee from Zimbabwe who has become a Cape Town taxi driver, explained to me how being able to spot a customer immediately can "enable you to have a wage for several months"; just as not spotting one can mean "putting your life on the line." The singular character of Upper Long Street has its origin in the heterogeneous nature of the voices which pass along it. Different needs, desires, and possibilities, different economic, psychological and social backgrounds all meet up and intersect here.

I often ended up heading for The Stones: a large billiard hall, which becomes a dance club on weekends. Customers can have a break on one of The Stone's balconies while they smoke a cigarette or have a drink and watch passersby down on the street. I would stay there for hours, observing the people walking along the street. Some were dancing, others drunk; sometimes I saw tourists walking by very quickly because they were afraid, while some drug dealers were trying to sell drugs to the passersby.

After a few months, I learned to recognize different characters who seemed like extras in a play to me. Mumford compared the city to a theater, "a collective experience in which dwellers are actors rather than merely spectators in the drama of urban life" ([14], p. 2). Long Street struck me as a theater stage whose habitués could play particular roles on it. In the Pretoria, a bar frequented by a predominantly black clientele, I got to know Jane, a prostitute from Mali, who would pick up her clients there. One day I saw her wearing a T-shirt with the slogan "No money: No honey." Like on a stage, Jane played the role of herself, taking on a particular role and giving different personalities to the other characters who were "acting" on her stage. She knew the street's every nook and cranny and would study the behavior of her potential clients who she had even subdivided into twelve categories: "There are clients who want to be with you for a night; these ones just want sex. There are those who want to stay with you for a few days; these ones want company. There are those who want a relationship; you can stay with those for months, even years..."

Sometimes I would go out onto the street and walk around listening to how the different types of music, sounds and voices emanating from the bars overlapped. For me this interwoven palimpsest of sounds brought to mind another of the city's polyphonies: those of the submerged voices, of the personal histories from the deepest levels, familiar to no one, but which become superimposed in the street through the meanings that their life experiences elicited. "To capture the hidden meanings," and "to reveal the invisible"—this is how I see the role of anthropology. I would have liked to capture all the meanings of Long Street, but the street was incommensurable and my view only partial. I decided to surrender myself to its polyphony.

After many months spent in the field, I had given up on the idea of finding an explicative synthesis and had started to accept the contradictions of the street. In order to capture the multiplicity of the voices of Long Street I had had to lose myself within it. By listening to the life stories of my interviewees and going through the places of the city with them, I had observed it from different points of view, adopting the perspective of the taxi driver, the young middleclass man out for a good time, the tourist, or the prostitute.

By observing Long Street, I have discovered how my fieldwork is made up of the multiplicity of the viewpoints of people whose lives follow different paths, who have different aspirations, dreams, desires and fears. Every time I attempted to find a "synthesis" for Long Street, to find a stable and definitive viewpoint to which a single meaning could be anchored, it crumbled before the multiplicity of the street's voices.

Long Street could not be considered as a single subject; there were an infinite number of Long Streets, as many as the points of view people had on it, and these changed continually depending on how it was regarded: the time of the day, everyday events, meetings and fleeting thoughts which constantly transformed the lives of my interviewees. Long Street could be observed only by accepting its heterogeneous nature. Long Street is marked by the continuous flow of different viewpoints, voices, behaviors, different ways of perceiving its urban spaces that do not converge in a single homogenizing perspective, branching instead into a polyphonic view. Bauman compares the identity of modern society to a prism through which different aspects of contemporary life are codified and understood ([15], p. 471). We can use the same metaphor to observe the identity of urban spaces: Long Street, in this instance. The street was ensnared in the paradox of being unique and multiple at the same time. Like a prism, though it was defined by a single organism, it could reflect the various identities that crossed through it.

I wondered how I could represent such enormity. The only way to represent Long Street in its entirety and totality would be to create a map representation that could include each building, establishment, alley, traffic light, and object in it and try to understand the meaning it had for each of its regulars. This solution was not only unrealistic, but would also be useless and fruitless.

The Argentine writer Jorge Luis Borges, in the short story "On Exactitude in Science" [16], envisions a realm where the art of cartography is so important that the map of a province covered the entire area of the city. In this story, Borges narrates the story of a cartographer who creates a map so vast and accurate that it covers the entire kingdom. However, later generations found this representation so useless and meaningless that they decided to leave it to the "Inclemencies of Sun and Winters" in the "deserts of the West." In this story, Borges poetically suggests that any attempt to represent a territory in its entirety is both useless and destined to fail. A map that tries to assimilate the territory which it is intended to represent in a total way, conceals the intention of negating its otherness; it is a delirium of omnipotence. No scientific or artistic representation, and especially no ethnographic representation, will ever coincide with the actual object. The attempt to assimilate the other (whether a person, a social group, a ritual, or an urban territory) in its totality entails the very negation of its essence. Emmanuel Levinas [17] noted that it was only through recognizing the otherness of the other that can we establish any kind of a relationship based on recognition. Discovering Long Street let me recognize it in its incommensurate, unrepresentable nature. Attempting to portray it completely meant more than venturing on an ill-advised path; it meant threatening its otherness. But if the map of Long Street, like any kind of map, can never grasp the territory in its entirety, which aspects should I consider and which should I ignore? And what point of view should I take to describe it?

3. Bateson at Cape Town

On this point, an important source of inspiration came from one of Gregory Bateson's early works, *Naven* (1958), in which he explores a ritual of sexual inversion of the Iatmul people of the Skip River in New Guinea). Bateson organizes his work into a series of chapters in which he observes the ritual through different analytical perspectives. *Naven* was initially met with very little attention, but later was reevaluated and became an anticipatory example of post-modern ethnography. George Marcus [18] cites it as an example of a text for the new alternative forms of ethnographic representation: "Bateson's experimental ethnography, which worries over several alternative analyses of a single ritual of a New Guinea tribe, is remarkable precisely because it was exceptional and unassimilated in the anthropological literature for such a long time, but now is an inspirational text in the current trend of experimentation" ([18], pp. 40–41). According to Marcus, what ties Bateson's anthropology with post-modern experimental trends is precisely its multi-perspective and multi-narrative approach: "One common idea of text construction is to string together a set of separate essays dealing with different themes of interpretation of the same subject" ([19], p. 192). James Clifford considers *Naven*'s methodological collage as a major forebear of the poetics and politics of cultural inventions. It can be considered an example of what he calls ethnographic surrealism: "The ethnography as a collage would leave manifest the constructivist procedures of ethnography knowledge: it would be an assemblage containing voices other than the ethnographer's, as well as an example of "found" evidence data not fully integrated within the work's governing interpretation" ([20], p. 147).

In many ways, *Naven* can be considered a precursor of major ideas that have recently informed the study of cartography and the practice of mapping in general. Over the last twenty years, interest in maps and the practice of "mapping" has grown in several disciplines, including anthropology, cultural studies, marketing, museum studies, architecture and popular music studies. Les Roberts [21] has emphasized how this interest can be traced to the impact of a epistemological reorientation, called a "spatial turn," in which scholars in diverse disciplines approached their field of interest with a flourishing "spatial critical imagination," finding its "defining trope" ([21], p. 18) in the use of maps. At the same time, disciplines like geography and cartography, which have traditionally used maps as analysis tools, have started to reconsider the very nature of maps as a tool of representing reality. This "cartographic turn" was bolstered by the publication of *Maps, Knowledge and Power* (1988) in which Brian Harley presented an analysis of the social nature of map representations with its connections to political power. The authors reconsider the idea—widespread since the Enlightenment—that maps are a neutral, objective representation of the world. He highlights that maps can only be understood in light of the historical, political and social context in which they were made. Maps, far from presenting "the truth of the world," were and are "social constructions presenting subjective versions of reality" ([22], p. 43). From this perspective, a map, any map, can only be read as a historical and social product situated in a specific context and in light of the cultural values driving its maker. According to Harley, maps act as metaphors for a "discourse" (meant as a system of meanings) that is more general than what they reference. The sense of the map as a "mirror of nature" ([22], p. 34) comes from the non-recognition of discourses (contained in them). The author considered the deconstruction of a map as a tactic of reading through which we can challenge "the apparent honesty of the image." ([23], p. 21) While it is true, in the words of Jeremy Crampton [24], that the practice of mapping has been continually contested, Harley's challenge to the presumed scientific objective of maps helped generate a more explicit critique of cartography, with renewed interest about the practice of cartography and mapping in general. Several scholars have studied the practice of mapping by researching the historical, political and ideological contexts in which the practice emerged and developed (Anderson [25], Cosgrove [26], Dodge, Kitchin, Perkins [27], Wood [28]). Like Bateson, the cartographic critics start from the assumption that no map can be considered a faithful replica of the territory, and every representation must be read and considered as a limited vantage point of the observer. The analytical convergence between Bateson and scholars of cartographic criticism leaves room for adopting models of multiple observation and representation of the same territory.

Drawing from these ideas, I envisioned a map of the city that could observe, describe and represent its inhabitants' different ways of perceiving the city. This map could not be considered a mere representation of the city, but rather an observation of how its inhabitants produce the territory. As Bateson said, the appreciation of differences is not just recognizing the traits of a territory or multiplying viewpoints on it, but it is the differences that arise and develop from the relationship between the two agents. Accordingly, in my map, I sought to identify the discrepancies in the different ways that its inhabitants relate to the territory, perceive it and relate it to their own socio-cultural situations and individual lives.

4. On Method

In building a map of Long Street, I drew from several sources both from the worlds of art and ethnographic research.

One of the most important sources of inspiration came French surrealists' deambulations. In surrealist thought, the act of wandering aimlessly is considered related to the practice of automatic writing, through which they sought to explore the spontaneous mechanisms of human thought. The French surrealists saw space as an empathic place (an active, living subject), with which they could forge a mutual dialog. The act of wandering can be seen, in this sense, as an attempt to explore a place's invisible reality, as well as a way to investigate unexplored facets of one's inner life (or ones that could not be explored otherwise). Inspired by surrealism, I sought to explore the memories, hopes and desires of my interviewees by wandering through the city. As in the surrealists' deambulations, I considered Cape Town's urban territory as an empathetic territory through which I could investigate the inner lives of its inhabitants. Wandering through the city's streets with the interviewees, I explored their memories, hopes and desires in relationship to the city's urban spaces.

Another essential contribution came from Andrew Irving's research in Kampala, London, and New York about how HIV-positive individuals perceived the urban territory (2004–2008). Irving's method involved identifying different places in the city that the interviewees considered important for their personal experience connected to the disease. He intends these places as performance stages where the most important moments of their experience can be explored and performed. Revisiting a house abandoned in the outskirts of Kampala lets an interviewee explore the moments when he likely contracted the HIV virus; returning to the room of a hospital evokes the moment when he learned that he had contracted the virus. Moving through the city's areas with his interviewees, Irving traces urban pathways intended as mnemonic and emotional itineraries. This type of urban movement can be understood as a form of mapping the territory through which the inhabitants attribute meanings to urban spaces in the light of their introspective journeys. Irving's walking fieldwork is not merely a tactic for gathering data from the urban territory; on the contrary, it can be seen as a way of making place. By wandering through the city and projecting their memories, fears, hopes and emotions on the urban spaces, Irving's interviewees create urban pathways through which the city's urban territory takes on different meanings. Though Irving never produced a map of his work, walking fieldwork can be considered a practice of mapping in which the city's territory is not only rediscovered but also continuously recreated by its inhabitants.

In recent studies related to cartography, we have seen new theoretical directions, among which we could include a post-representationalist perspective in which the relationship between map and territory is less clear and delineated than it had been previously. Several authors (James Corner [3], Tim Ingold [8], Rob Kitchin and Martin Dodge [27], Vincent Del Casino and Stephen Hanna [29]) have contrasted the static, stable conception of the structural school, favoring a dynamic conception based on peformativity in which the map is seen a process, or rather a "constellation of ongoing process" ([27], p. 547). Martin Dodge, Rob Kitchin and Chris Perkins have noted that this approach reflects "a philosophical shift towards performance and mobility and away from essence and material stability" ([27], p. 554). The map stops being considered a fixed, stable product and is seen as something in a state of becoming. Vincent Del Casino and Stephen Hanna see the map as a flux in which each encounter produces new meanings and engagements with the world. "Maps," they write, "are not simply representations of particular contexts, space and time" ([29], p. 42). On the contrary, "they are mobile subjects infused with meaning through contested, complex, inter-textual and interrelated sites of social-spatial practices." ([29], p. 42). The map is a process in a constant state of becoming, created through a series of practices that are reflected in how the territory is understood. "It can be seen as, far from fixed, unmoving entities" ([29], p. 45) In a study in the historical Town of Fredericksburg in Virginia, they explored how visitors to the town were able to enrich the meaning of tourist maps through contact with the territory, and how, at the same time, the map was a tool through which they could give a particular meaning to the town's places. Map and territory are understood as co-constituting agents to the extent that we cannot recognize how much the map is built by the territory and the territory by the map. Map and territory can be understood only in the light of the complex, interdependent relationship between them. Comparably, Rob Kitchin and Martin Dodge emphasized how maps cannot be understood as "secure representations" but rather as "a set of unfolding practices" ([27], p. 557). Maps are transitory and fleeting, being contingent, relational and context dependent" ([27], p. 559).

Drawing from these works, we can envision building a map based on urban movement that aims to represent a city through the different ways its inhabitants have of producing the territory. Put differently, we can imagine a map of the city made up of different urban pathways chosen and built by the city's inhabitants by wandering in the city and which do not just represent the territory but recreate it. Borrowing from the French surrealists' deambulations and Irving's walking fieldwork technique, I built a map of the city through the urban movement of its inhabitants. However, in my case, urban movement was seen as more than a tactic for exploring the relationship between my interviewees' inner worlds and the city's urban spaces; it had the ultimate aim of revealing how the relation to different social groups, economic conditions and personal experiences led the people of Cape Town to map the urban territory differently.

At the beginning, I chose ten regulars of Long Street who had different traits in terms of gender, language, race, political orientation, economic level and socio-cultural background. I then asked my interviewees to choose ten places in the city that had particular importance for them because of being (directly or indirectly) tied to their personal experiences. Next, I asked my interviewees to move through the city's places, re-evoking the feelings and impressions that were elicited. These places were seen as mnemonic and emotional areas in which they remembered and evoked important memories from their pasts. While taking the urban pathways I photographed the places, recorded the interviews and wrote down their comments. In the final phase, I asked my interviewees to explore the pathways we took, showing their routes drawn on a sheet of paper and the pictures taken on the pathways, taking notes of the comments and memories aroused by this phase of map and photo eliciting.

The first step entailed tracing lines to mark the pathways that I had taken in the city by myself and with my interviewees, drawing the places where we did the interviews. I personally drew the territorial stages, taking inspiration from a photographic map of the city. The drawings are intentionally approximate and do not faithfully reflect these areas (Figure 1). My goal was to represent the "distance" between my topographic representation and the urban territory.

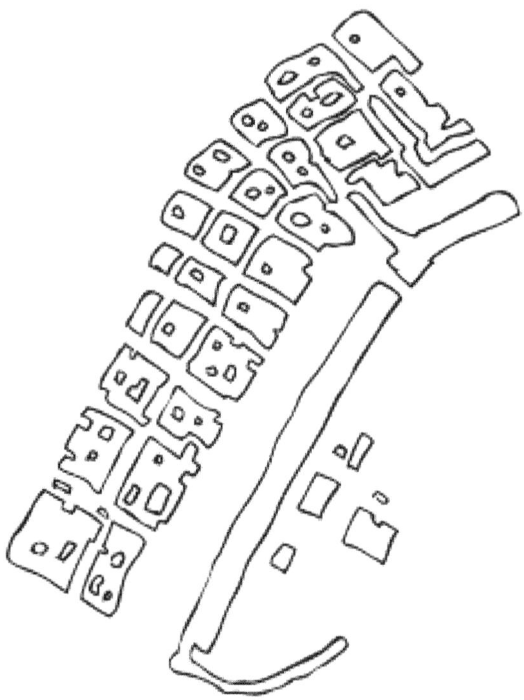

Figure 1. Long Street.

In a second phase, I included writings on my map taken from conversations I had with my interviewees and put them in relationship to the urban pathways taken. The writings serve to represent the urban territory through the meanings that its people attribute to them. By aligning with writings taken from interviews with the drawings on the map, a meaning can be attributed to them. The second function of the writings is to orient visitors in discovering the city. By relating the writings with the lines drawn on the map, readers can go over the urban pathways taken during the interview and match them with the interviewees' mnemonic reconstructions.

For example, in an interview with Kay, a 35-year-old black woman originally from a town in Western Cape who came to Cape Town when she was 20, I traced the most important moments of her experience in the city. She remembered the period when she was involved with a Russian man and told me how the fact that she was going out with a "white man" let her access a world from which she felt excluded. Walking by Camps Bay, she remembered a day when she went to a restaurant with her boyfriend and was served by a white waitress for the first time in her life. She said:

> "One night we came here and there was a white waitress serving the tables. When I saw her I felt a sense of triumph. I'd spent many years of my life listening to white customers' protests and complaints, often while keeping a forced smile on my face. Now I was on the other side of the fence. I went into the restaurant, she greeted me and I didn't reply and put my arm around my partner: it felt fantastic."

Connecting a phrase from an interview with Kay to the representation of Camps Bay, I started to map the pathway taken with her (Figure 2).

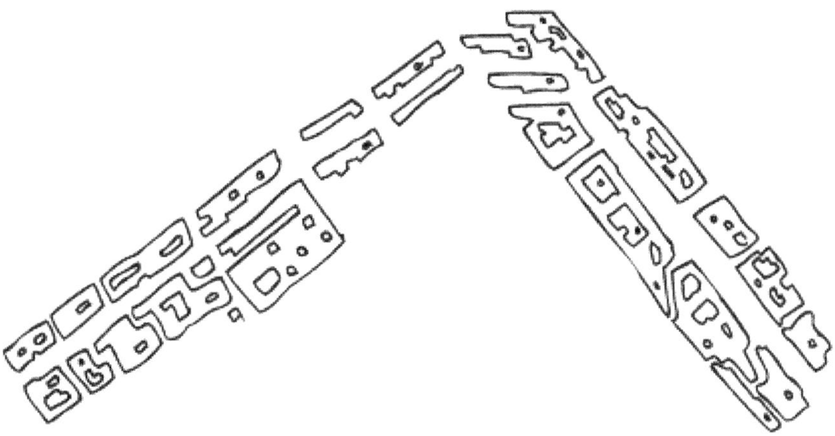

Figure 2. "Now I was on the other side of the fence."

In an interview conducted with Ibrahim, a young Somali man who came to Cape Town in search of a city where he could live freely as a homosexual, I went to the club where he had his first job and where he felt free to express himself for the first time.

> "I've found this work thanks to a friend. A foreigner guy who had many contacts in the gay environment of Cape Town. When he told me that I had a job, I could not believe that. When I cross the door of the club it was like passing the finishing line of a long walk that I did for years. Here things have changed completely. I've started dressing and speaking as I like. Acting and pretending are what had characterized my whole life. Acting, playing a part was a pleasant feeling, because I could maintain a secret identity, I could cherish it and protect it. In a certain sense, acting gave me the freedom to be

myself in the purest sense, because the mask I put on protected me. Coming here brought about a transformation in my life and in a certain sense I've stopped acting. When I started working here, I discovered another part of myself. It was like I was hiding all over again (Figure 3)."

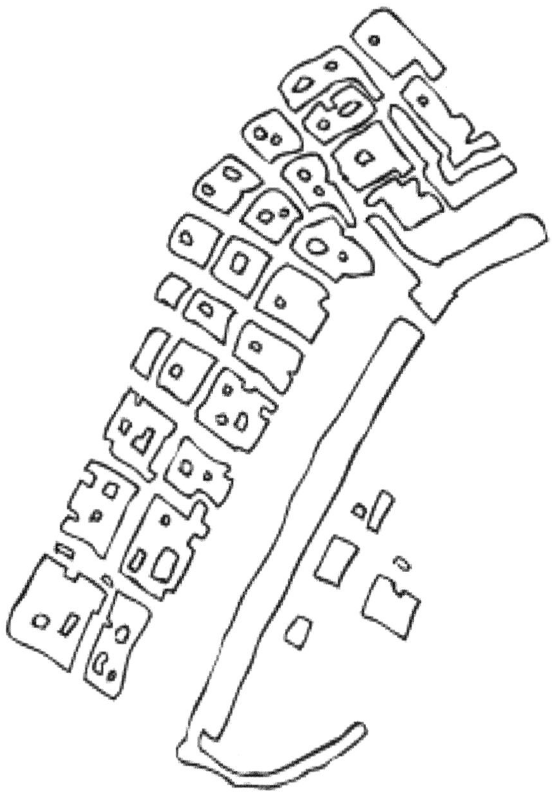

Figure 3. "When I cross the door of the club it was like passing the finishing line of a long walk that I did for years."

In a third phase, I connected the territorial stages through some vectors representing the urban pathways taken with my interviewees. The vectors suggest a possible movement, but leave open the option of taking other pathways. Recalling *The Naked City* [30] by Guy Debord, the city is intended as a psychological landscape comprised of holes. Entire parts are forgotten or erased to build entire possible cities in the empty space. The model is an archipelago, a series of islands surrounded by an empty sea cut through by wanderings (Figure 4). The empty space that separates the territorial stages underscores the possibility of taking new urban pathways and developing new interpretations. The map is simultaneously an invitation to take the pathways already marked and an invitation to get lost amidst its different ethnographic landscapes.

Figure 4. Ibrahim pathway.

A third element of the map is made up of pictures portraying where we went and taken in the pathways. The function of these pictures in an exploratory sense was both to orientate a discovery of the city, establishing points of reference, and evoking memories in the interviewees in the photo elicitation phase. Likewise, placing the photographs (e.g., Figures 5 and 6) on the city map, I sought to guide readers to discover the city, showing the places that we passed through during the research phase. By putting the photographs into relief with the writings and drawings, readers can reconstruct the urban pathways.

The last step for building the map was assembling the data. In this phase, I thought of how to put together the elements that I had collected in my work and how I could represent Long Street.

A suggestion on how to think about this came out of a happenstance visit to the District Six Museum in Cape Town and the discovery of the district's map. The memorializing District Six Museum is one of the museums in Cape Town that is most recognized internationally. The aim of the museum is to promote social justice "as a space for reflection and contemplation, and as an institution for challenging the distortions and half-truths, which propped up the history of Cape Town and South Africa." ([31], p. 152).

Figure 5. Kick Ass on Long Street; interview with Ibrahim.

Figure 6. Camps Bay interview with Kay.

The District Six Museum Foundation drove the museum's creation. The organization was founded in the late 1980s with the mission of preserving the memory of the district's ex-residents in the belief that this could impede the rise of new forms of social oppression. One of the foundation's first initiatives was in organizing the exhibition *Streets* (1994); it sought to represent District Six through street signs that had once appeared in the neighborhood. Sandra Prosalendis, the director and founder of the museum, remembered when she was trying to gather the material for the first exhibitions. She explained that one idea about how to arrange the street signs was to put them back in their original places, but this idea was soon abandoned: "The landscape as it exists, as a space of absence, has its own message; it is a huge scar in the center of the city, and it speaks volumes" ([32], p. 59). The museum's foundation decided to display District Six's urban area as it was before the demolitions, represented through a map created by several artists, including many ex-residents of the district. Prosalendis enthusiastically remembers when the exhibition first opened:

> "From the minute the museum opened its door to the public, the ex-residents came and began to sign the map themselves, to write themselves back into the center of the city, to claim their history, and to claim the space. People marked bus stops, places where somebody sold peanuts, their old schools, and their homes." ([32], p. 61).

Annie Coombes noted how the act of writing on the map and the other museum installations served to establish a direct dialog "without mediation" between the ex-residents and its visitors. By inscribing the district map with their thoughts, hopes, memories and dreams, District Six's ex-residents sought to give voice to the district's "dehumanized" places. The District Six map still covers almost the entire surface of the museum's first floor and is used by the guides and museum attendants to describe the district's history and various changes. The work in itself may look like a regular representation of the district, not unlike the anonymous topographic representations made by apartheid's planners: "It is empty, devoid of life, able to be manipulated in the interests of those in authority" ([33], p. 143). It shows the arrangement of the streets and their times that once defined the urban district. However, it takes on a special meaning when we walk on its surface. Walking on the map's surface and beyond its borders, we can explore the different meanings that its residents attributed to its urban places. On the surface of the installation and those around it, there are sheets, molds and objects imprinted with the ex-residents' writings. Visitors discover through the map the different meanings once attributed to the district. In this sense, the District Six map should not be seen only as a representation of a territory, but a territory itself that can be explored by crossing it. Charmaine McEachern noted that the District Sixers used exactly the same representation of the district during apartheid, which on the floor of the museum also began as an "empty" representation but took on a completely different meaning: "But their articulation of memory and walking provide for it a totally different meaning, one which resists the apartheid regime's judgment, while at the same time criticizing its acts of destruction" ([34], p. 505). Walking on the map allows us to discover that territory through the multiple meanings that its residents attributed to its spaces (Figure 7). In this sense, walking on the map can be understood as "making place" in which the urban territory's spaces take on meaning and substance. While it is true that by building the District Six map the ex-residents of the district recreate the urban territory by projecting onto it their thoughts, memories, hopes and desires, it is by moving on the map that we can discover them and bring them back to life.

Figure 7. The District Six Museum walking map.

The practice of walking can be seen as a place making through which the reader "appropriates" the territory through the meanings that its ex-residents attributed to it. The map does more than show the urban movement on the territory; it provokes new types of movement with which to explore the city. Reading the cartographic representation of the city and the practice of mapping as construction of the territory seem to coincide here. Taking inspiration from the District Six map and comparing it with Bateson's epistemology of differences, we can envision a kind of ethnographic representation of the city that is both map and territory, and where its differences can be discovered through movement. Based on these premises, I started to imagine a cartographic representation of the city made up of different meanings that its inhabitants attribute to its urban places and in which the action of walking coincides with their interpretation. Like the District Six map, I would create a platform from which the city could be explored through the act of walking and where it could be discovered through its differences.

By combining the three constituent elements of the map, the map drawings, photographs and written texts, I sought to create a multiple perspective representation of the city. "Long Street" is made up of different forms of communication whereby the interactions create a hybrid, multi-dimensional representative form. The drawn maps, the written text and the visual texts are not meant here as an illustrating support for each other, but as complementary, integral parts of one expressive, interpretive flow. Inspired by the District Six map, I decided to think of my map as a walking map through which visitors could explore the street. Placing the territorial stages, vectors, writings and photographs on the platform (through paper representations and/or projections), I created a map-territory of the city. Walking on the map of Long Street, the city is discovered through the different perspectives of its residents. Walking on the map is intended as a way of connecting the different viewpoints. However, this connection is precarious and feeble. We can change the pathway for the interpretation of the city to take on a completely different meaning. In this sense, the representation of Cape Town is multiple, both because the city is articulated through different voices and different languages. The combination of different languages helps to reveal the object's multi-dimensional nature. The map is multiple both in the object to which it refers and in the method that it uses.

5. Conclusions

In this article, I described the construction of an ethnographic map of Long Street representing post-apartheid Cape Town through the different viewpoints and voices of its inhabitants. The primary goal of this work was to observe the city through these multiple perspectives. My intent was to take different viewpoints on my object of study to create a crisis of belief in the viewer accustomed to reading an ethnographic account of the city as the only possible view of the city. In other words, I wanted to show that no map is ever the territory and that viewpoints and representations could be multiplied on the same territory, achieving different results. In order to expose the difference between my representation and the object of my study, I began to recognize the city as a subject. I had to listen to its different messages and voices without the fixation of having to bend them to a unified, all-encompassing view, whether holistic or reductionist.

For this purpose, a solution came from Bateson's epistemology of difference which he continued from his earliest works to his last ideas on the process towards an ecology of mind. Taking this epistemological perspective, the city could no longer be considered an object of study, but rather an animate subject that expresses itself through different voices. Long Street regulars were not considered "inhabitant objects" of post-apartheid Cape Town, cut off from the historic flow dictated by grand narrativees and political upheavals. Rather, they are "inhabitant subjects" who can shape, invent, and create their stories. In this view, the city is read as the whole of the inventions and creations of the memories, reasoning, and imaginations of its inhabitants. Because I could no longer give a final, all-encompassing explanation of the city, stopped at offering a montage of its voices so that they can continue to express themselves freely.

I considered the activity of "montage" not as something which follows from observation, but as a fundamental part of observation. Putting together the data, reflecting on it, discovering "new juxtapositions" between the life stories I had listened to and the places I had visited could not be considered merely a stylistic exercise, as it was also an integral part of its hermeneutic activity. In this phase, I also found a convergence between Bateson's ideas and the District Six map. George Marcus noted that, for Bateson, the method was expressed less in what was done in the field than in what was done with the data when behind the desk ([18], p. 295). It is the preparation of the data, not its collection, that creates the method. In *Naven*, the expression of contrasting methods is meant as the premise of their combination through montage. The explanation is therefore a collage of different perspectives. Likewise, it is through the assembly of different elements that the District Six map constructs its narrative to represent the district. We saw how the creation of the museum's map (and the museum itself) was shaped by juxtaposing different elements of the district, such as street signs, objects and clothes that the ex-residents brought to bear witness. We saw how its reconstruction (through the assembly of these elements) can be seen as bringing the district back to life. The map is not, as such, a faithful replication of District Six, but a (political and poetic) remixing done through a new order suggested by its residents. District Six is represented in the museum as a district in fragments, reconstructed through its material and immaterial elements and waiting to be reconstructed through the visitor's gaze. For Bateson, representation is not the attempt to replicate the object of study, but a process of deconstruction (destruction) and reconstruction of the data. We can see how the way that its ex-residents reconstruct District Six cannot be reduced to a mere exercise in style, but should be tied to a particular way of historically interpreting apartheid.

Annie Coombes noted that one of the District Six Museum's distinctive traits is in replacing apartheid's monologic perspective with a dizzying multitude of different voices. The homogeneous, internally cohesive apartheid ideology is replaced by the heterogeneous collection of its citizens' thoughts, reflections, doubts, and hopes. We can follow the museum's trajectory only by taking on the different viewpoints, thoughts, and memories of the ex-residents reconstructed through a series of installations, objects, and writings spread throughout the museum. Visiting the District Six Museum, visitors are invited to lose themselves amidst the viewpoints of the district's ex-residents and to adopt a multiple, differentiated viewpoint. Every point of view of its inhabitants can be read and interpreted through the social and cultural product of its origin, but they do not converge. Apartheid's monologic perspective is answered with a response of overlapping voices. Considering the map in particular, we can see how the act of walking coincides with discovering the district through its differences. It is movement on the "territory/map" that reveals the district through its polyphony. This map of Long Street does not seek to represent Cape Town through a single point of view; instead, it is an invitation to get lost in the different ways of observing the city. This invitation to walk on the map of Long Street is to be taken as an invitation to build one's own interpretative pathway and one's own map of the city.

Conflicts of Interest

The author declares no conflict of interest

References

1. Gregory Bateson. *Steps to an Ecology of Mind: Collected Essays in Anthropology, Psychiatry, Evolution, and Epistemology*. Chicago: University of Chicago Press, 1972, pp. 465–67.
2. Alfred Korzybski. *Science and Sanity: An Introduction to Non-Aristotelian Systems and General Semantics*, 5th ed. Brooklyn: Institute of General Semantics, 1994, p. 35.
3. James Corner. "The Agency of Mapping, Speculation Critique Invention." In *Mappings*. Edited by Denis Cosgrove. London: Reaktion books, 1999, pp. 228–13.
4. John Allen. "Worlds within Cities" In *City Worlds*. Edited by Doreen Massey, John Allen and Steve Pile. New York: Routledge, 1999, p. 53.
5. Gregory Bateson. *Naven: A Survey of the Problems Suggested by a Composite Picture of the Culture of a New Guinea Tribe Drawn from Three Points of View*. Stanford: Stanford University Press, 1958.

6. Vincent Crapanzano. *Imaginative Horizons.* Chicago: University of Chicago Press, 2004.
7. Michel De Certeau. *The Practice of Everyday Life.* Berkeley: University of California Press, 1984.
8. Tim Ingold. *The Perception of the Environment: Essays on Livelihood, Dwelling and Skill.* London: Routledge, 2000.
9. Andrew Irving. "Life Made Strange: An Essay on the Reinhabitation of Bodies and Landscapes." In *Qualities of Time: ASA Monograph.* Edited by Wendy James and David Mills. Oxford and New York: Berg, 2005, pp. 317–30, vol. 41.
10. Andrew Irving. "Journey to the End of the Night: Disillusion and Derangement among the Senses." *Journeys* 9 (2008): 138–60.
11. Sean Field, Renate Meyer, and Felicity Swanson. *Imagining the City.* Cape Town: HRSC Press, 2007, pp. 6, 7, 12.
12. Noëleen Murray. *Desire Lines, Space Memory and Identity in the Post Apartheid City.* London: Routlege, 2007, pp. 8–9.
13. Achille Mbembe, and Sarah Nuttal. "Writing the World from an African Metropolis." *Public Culture* 16 (2004): 347–72.
14. Peter Critchely. *Lewis Mumford and the Search for the Harmonious City* (e-book). 2004. Available online: http://independent.academia. Edu/PeterCritchley/Books (accessed on 18 September 2013).
15. Zygmunt Bauman. "Identity in the Globalising World." *Social Anthropology* 9 (2001): 121–29.
16. Jorge Louis Borges. *A Universal History of Infamy.* London: Penguin Books, 1975.
17. Emanuel Levinas. *Totality and Infinity: An Essay on Exteriority.* Pittsburgh: Duquesne University Press, 1961.
18. George Marcus. "A Timely Reading of Naven: Gregory Bateson as Oracular Essayist." *Representations* 12 (1985): 66–82.
19. George Marcus, and James Clifford. *Writing Culture the Poetic and Politics of Ethnography.* Berkeley and London: University of California Press, 1986, p. 192.
20. James Clifford. *The Predicament of Culture: Twentieth Century Ethnography, Literature and Art.* Cambridge: Harvard University Press, 1988, p. 147.
21. Les Roberts. *Mapping Cultures: Place, Practice, Performance.* New York: Palgrave Macmillan, 2012, p. 19.
22. John Brian Harley. "Maps, Knowledge and Power." In *The Iconography of Landscape: Essays in the Symbolic Representation, Design and Use of the Past Environments.* Edited by Denis Cosgrove and Stephen Daniels. Cambridge: Cambridge University Press, 1988, pp. 277–312.
23. John Brian Harley. "Deconstructing the Map." *Cartographica* 26 (1989): 1–20.
24. Jeremy Crampton. *The Political Mapping of Cyberspace.* Edinburgh: Edinburg University Press, 2003.
25. Ben Anderson, and Paul Harrison. "Commentary, Questioning affect and emotion." *Area* 38 (2006): 333–35.
26. Denis Cosgrove. *Mappings.* London: Rethinking books, 1999.

27. Martin Dodge, Rob Kitchin, and Chris Perkins. *Rethinking Maps: New Frontiers in Cartographic Theory*. London: Routledge, 2009.
28. Denis Wood. *The History of Cartoghraphy—Volume 1 Cartography in Prehistoric, Ancient and Medieval Europe and the Mediterranean.* Edited by John Brian Harley and David Woodward. Chicago and London: University of Chicago Press, 1987, p. 659.
29. Vincent Del Casino, and Stephen Hanna. "Beyond the binaries: A methodological intervention for interrogationg maps and representational pratices." *ACME An International E-journal for Critical Geographies* 4 (2005): 36–56.
30. Guy Debord. "The Naked City." In *The Situationist International: A User's Guide*. Edited by Lewis S. Ford. London: Black Dog Publishing, 1958.
31. Rodney Harrison. *Heritage Critical Approaches.* Abington: Routledge, 2013.
32. Valmont Layne. "The District Six Museum: An ordinary people's place." *The Public Historian* 30 (2008): 53–62.
33. Annie Coombes. *History after Apartheid: Visual Culture and Public Memory in a Democratic South Africa*. Johannesburg: Wits University Press, 2003.
34. Charmaine McEachern. "Mapping the Memories: Politics, Place and Identity in the District Six Museum, Cape Town." *Social Identities* 4 (1998): 457–78.

MDPI AG
Klybeckstrasse 64
4057 Basel, Switzerland
Tel. +41 61 683 77 34
Fax +41 61 302 89 18
http://www.mdpi.com/

Humanities Editorial Office
E-mail: humanities@mdpi.com
http://www.mdpi.com/journal/humanities

www.ingramcontent.com/pod-product-compliance
Lightning Source LLC
Chambersburg PA
CBHW040810250926
51850CB00007BA/1127
* 9 7 8 3 0 3 8 4 2 1 6 5 8 *